Never Lose Your Nerve!

Never Lose Your Nerve!

Alan J. Heeger

University of California, Santa Barbara, USA

 **World Scientific**

NEW JERSEY · LONDON · SINGAPORE · BEIJING · SHANGHAI · HONG KONG · TAIPEI · CHENNAI · TOKYO

Published by

World Scientific Publishing Co. Pte. Ltd.

5 Toh Tuck Link, Singapore 596224

USA office: 27 Warren Street, Suite 401-402, Hackensack, NJ 07601

UK office: 57 Shelton Street, Covent Garden, London WC2H 9HE

British Library Cataloguing-in-Publication Data
A catalogue record for this book is available from the British Library.

NEVER LOSE YOUR NERVE!

ISBN 978-981-4704-85-4
ISBN 978-981-4704-86-1 (pbk)

Printed in Singapore by Mainland Press Pte Ltd.

Never Lose Your Nerve!

Alan J. Heeger

This Book is dedicated to Ruth, who "has filled my life with love and surrounded me with beauty" and made all things that were impossible — possible.

Prologue

Professor Heeger shared the Nobel Prize in Chemistry in the year 2000 with his two colleagues, Hideki Shirakawa and Alan MacDiarmid, "for the discovery and development of conducting polymers." This book is the story of his path from a small town in Iowa to receiving the Gold Medal that is the symbol of the Nobel Prize in Chemistry.

It is not one story but a set of intertwined stories, parallel in time, that describe a rich and complex life: a love story; a description of his passion for science and the various stages from student to young Professor to the award of the Nobel Prize; his love of and involvement with the theater and how that involvement had direct impact on his science; his adventures as an entrepreneur; and his love for and his impact on his two sons, Peter and David. Throughout this quiltwork of love, passion for science, entrepreneurship, theater and the award of the Nobel Prize, he expresses a love of the mountains and of skiing, which he uses to illustrate many of the significant events of his life.

He tells us of a "miracle" that actually happened during his childhood.

The loss of his father when he was nine years old shaped his life and forced him to accept his own mortality at this early age. Consequently, his mother guided and pushed him toward advanced education all the way to the award of the PhD degree in Physics from the University of California at Berkeley.

His love for his wife, Ruth, provides the foundation for the entire tale. "Ruth has filled my life with love and surrounded me with beauty. Throughout the 64 years that we have been together, our love carried us very far. Our love is responsible for what we have become. Our love and Ruth's support are fundamentally responsible for all of our accomplishments including my success as a scientist, the award of the Nobel Prize and the becoming of a successful entrepreneur."

Although he received his PhD at Berkeley, he became a Scientist at the University of Pennsylvania, where he was appointed as Assistant Professor in 1962, and rapidly promoted to Associate Professor and then to Full Professor in 1967 — at that time the youngest Full Professor in the Ivy League Universities.

During this 20-year period at Penn, he developed a close relationship with his colleague, Nobel Laureate J. Robert Schrieffer. Robert Schrieffer is a great man of science who became famous at the tender age of 25 when he created in his mind the macroscopic, many-electron wave function that describes the remarkable phenomenon known as superconductivity. Superconductivity, discovered in 1911, had remained a mystery and the principal unsolved problem in Solid State Physics — the BIG Problem — for nearly 50 years until 1956, when John Bardeen, Leon Cooper and Robert Schrieffer published their BCS Theory of superconductivity. In the early 1960s, Schrieffer was "hot property"; he was sought after and recruited by the physics departments of many major research universities across the United States, including the physics department at Penn. The excitement of the opportunity to build a great Physics Department convinced Bob Schrieffer to come to Penn as the Mary Amanda Wood Professor of Physics.

Professor Heeger completed his education as a Physicist with Schrieffer as his mentor. They did not actually collaborate on a specific research project or publish together as co-authors until more than 15 years later. During the early to mid-1960s, Alan learned from Bob to aspire to achieving and creating important physics. "In order to become a great scientist, one must work with and learn from a great scientist, not just to learn the facts — the facts can be learned from books — but to learn what is truly important to the progress of science. I learned these in my early years at Penn from Schrieffer."

Perhaps the greatest pleasure of being a scientist is to have an abstract idea, then to do an experiment (more often a series of experiments is required) that demonstrates that the idea was correct; that is, Nature actually behaves as conceived in the mind of the scientist. This process is the essence of creativity in science. Professor Heeger is fortunate to have experienced this intense pleasure many times in his life.

The underlying theme of the book is the role of Creativity, Discovery and Risk in science. People typically think of scientists as meticulous and focused, perhaps even boring. Many think that scientists do not tolerate risks. Although this is true for some scientists, those that are risk-averse are not the creative and productive scientific leaders. "In fact, for me and for most scientists, risk-taking is part of our lives; we are risk-addicted. Every time we publish an article, we take on risk. We try to make certain that the data are correct, but it is impossible to be certain. We seriously try to give the correct interpretation of the data. But the very process of research involves pushing beyond what was previously known; that process involves taking risks. Scientists take risks all the time. And, of course, the more interesting the result, the bigger the associated risk."

Interdisciplinary science is even more risky. Educated as a physicist, Heeger had a core of knowledge with which he felt very comfortable. Each time he reached out beyond that core, he was exposing his ignorance. But reaching out into new directions involves learning new concepts and finding a way to meld those new concepts into what one had known previously. Exploring new directions is the first step toward creativity and discovery. Nevertheless, dealing with that risk is absolutely essential: ***One must never lose one's nerve!***

After coming to the University of California at Santa Barbara, he maintained his world leadership role in interdisciplinary science, co-founded the Materials Department, now considered to be one of the best Materials Departments in the world, and also caught the "California Entrepreneur" disease. Heeger's first company, UNIAX Corporation, was co-founded with his colleague Paul Smith in 1990. UNIAX succeeded and was acquired by DuPont in the year 2000. He subsequently co-founded two biotech companies, both of which currently offer exciting promise for approaches to personalized medicine and the diagnosis and cure of cancer.

He presents an insightful discussion of the delicacy of the creative mind. His own firsthand experience with depression and subsequent recovery are described in the context of the many examples of great scientists who have suffered similarly — all too often with more serious consequences.

He concludes by noting that "the progress in science will go on long after I am gone," and therefore provides some advice for young scientists who aspire to scientific accomplishments that could have sufficient impact to generate nominations and eventually to result in the award of a future Nobel Prize.

To summarize, this is a book with content that goes well beyond the particular life of the author. It is filled with humor, with success, and with the sadness resulting from the early death of his father and — many years later — with the devastating sadness of losing the only person he truly loved other than his wife. It provides one with a view of life as an exciting adventure that can be a guideline for any creative person on the track of major impact on the lives of people now and in the future.

Acknowledgements

I acknowledge important suggestions and comments from my wife, Ruth, from my son, Peter, and from my brother, Gerald. Incorporating their suggestions greatly improved the "story" and the quality of the book. I thank Nick Colaneri for introducing me to the story of the Caliph of Cordova. I am forever grateful to Professor William Salaneck (Professor of Physics at Linköping University, Sweden) for his active support of the award of the Nobel Prize in Chemistry to Heeger, MacDiarmid and Shirakawa.

Alan J. Heeger
March 2015

Contents

1 Perfect Days – and Some Not so Perfect

In the 11th century, when the Moors were in Spain, the Caliph of Cordoba was one of the most powerful men in the world. As an old man, the Caliph wrote his memoir — as old men are prone to do — and in that memoir he commented that although he had had a good life, he had during his lifetime only 14 "perfect days." After learning of the Caliph's 14 perfect days, I awoke in the middle of the night. I am basically and fundamentally a competitor; surely I could list more than 14 perfect days in my life. I immediately started to make a mental list.

I have enjoyed many important and memorable days, such as our wedding day, the days on which my sons were born, the day I defended my thesis and completed the requirements for the PhD and the day I received my first invitation from the Nobel Prize Committee to make a nomination for the Nobel Prize in Physics. These, and many more, were important and memorable, but not "perfect" because each involved new responsibilities, and new responsibilities come with worries and anxieties. As I attempted to list the "perfect days," it quickly became clear that many of them were spent in the mountains, on the ski slopes, at places such as Zermatt, Verbier, Morgins, Val d'Isère, Aspen, Vail, Jackson Hole, Keystone, Snowbird, Solitude, Alta and Park City. For me, skiing is the perfect vacation. The responsibilities and associated worries and anxieties are left behind; one must focus on navigating the steep slopes and doing so with grace that comes only with continuing practice that leads to skill. Moreover, the remarkable beauty of the high mountain terrain is intoxicating. Skiing is my only sport (I do not play tennis or golf). I exercise throughout the year with the goal of being in shape for the next season.

Skiing started for me when I was 32 years old. On a dreary winter Friday afternoon in 1968, I was determined to wrap up the final details of a manuscript for publication in the *Physical Review* before going home for the weekend. By late afternoon, I managed to have everything done with the required number of copies of the manuscript plus an appropriate cover letter in a large stamped envelope ready to mail. I got into my car and dropped the envelope into the mail box at the post office near the 30th Street train station in Philadelphia. On the way home I decided that we should get away for the weekend.

Someone had told me about a small resort hotel near Lancaster, PA; the food was reputed to be plentiful and good, and there were a variety of diversions available to the guests. I called and made the room reservation. Ruth and the boys were excited at the unexpected prospect of the fun weekend ahead. We quickly packed the necessities and departed on the one-and-one-half hour trip to our destination. As it turned out, the food was indeed plentiful, but not especially good. The weekend was memorable for only one reason: It was my first time on skis.

The single skiing hill was literally tiny; I doubt that it was even 30–40 yards in length, and the slope was gentle with a total vertical drop of perhaps 20 feet. Instead of a chair lift, the proprietors had installed a motor-driven rope tow. One grabbed hold of the moving rope and was pulled slowly back up the slope. Ruth and I rented boots and skis, and sleds were rented for the boys. Looking down the little slope, the sleds were to go on the left side and the "skiers" were instructed to go on the right side.

I fell nearly every time. I broke my glasses, blisters appeared on my hands because of improper use of the rope tow, and before the afternoon was over every muscle in my body ached — but I was hooked! Perhaps this, too, is associated with my "risk addiction." This unplanned and unexpected introduction to skiing was fortuitous, because we were scheduled to leave in a few months for a sabbatical year in Switzerland. There are few better places in the world to ski.

The year in Geneva was a special time in our lives. We made lasting friendships, and we thoroughly enjoyed the beauty of Switzerland,

Each of us learned to speak a little French, although to be honest, none became even close to fluent. Ruth succeeded in obtaining what she wanted by using the phrase "quelque chose comme ça" followed by pointing or miming as in the game of Charades. Our sons have long ago lost most of the French that they learned while living in Geneva, but they have retained authentic accents. I was never afraid to speak, but often had difficulty in understanding the answers. In a group social conversation, I was always two sentences behind. It is difficult to make insightful comments when you are two sentences behind.

I gave a series of lectures at the Physics Institute of the University of Geneva, and I had the time to complete a major review article. Let there be no mistake, however — the lasting memories from our year in Switzerland are built around our skiing activities in the beautiful Swiss mountains.

I shared an office in the Physics Institute with Bernard Giovannini. Bernard had spent a post-doctoral year in the Physics Department at Penn before moving into his permanent position as a young professor at the University of Geneva. Bernard spent his entire career at the University of Geneva, including a period as Rector of the university. He is now enjoying his retirement living on his "farm" in France near the Swiss border where he dedicates himself primarily to his music.

Upon arrival in Geneva, Ruth and I renewed our friendship with Bernard and his wife, Maria Rosa. Early in the Fall, long before there was even a hint of snow, my enthusiasm for the coming ski season resulted in a plan to share a chalet in Zermatt with Bernard and Maria Rosa for 10 days during the Christmas holidays. Bernard made the arrangements. He showed us the location of the chalet on a map of the area; it was part way up the mountain on a road which gave easy access, on skis, into the town of Zermatt.

We carefully researched the sporting goods stores to decide where to purchase our skis and equipment. In those ancient times, the skis were made of wood; the remarkable materials technology that makes the skis of today possible had not yet been invented. The tradition in Switzerland was to use very long skis; short skis, we were told, were unstable. To determine the length of my first pair of skis, I was instructed to stand up tall and extend my right arm up as high as I could. I was very proud of that first pair; they were heavy and solid with a length of 210 cm. The ski boots were made of leather and fastened with shoestrings more or less like the boots that one would wear on city streets during the winter. The bindings comprised a steel cable that was pulled tight around the back of the boot and fastened onto the skis in front of the boot by a spring-loaded mechanism. *The bindings were not safe.* Fortunately, none of us have been seriously hurt during our lifetime of skiing, although Ruth did have a close call when we skied at Les Diablerets.

In the years that followed, the introduction of high technology materials has radically changed ski equipment. Today, as an expert skier, my skis are much lighter weight and only 160 cm in length — nearly a six inches shorter on either end than my first pair! Skis are now "shaped" to make turning easier. The stiff boots, made from engineering plastic and fastened with a series of metal buckles, enable the skier to transmit fine adjustments from the knees and hips to the sharp edges on the skis, and the bindings "pop" open when subjected to stress beyond a predetermined level. *The bindings are now safe.*

Weeks prior to our departure, I began to check the newspaper every day for reports of the snow conditions at Zermatt. The news was not good; there had been some new snow, but relatively little. I was concerned that we would not have sufficient snow to enjoy our first ski holiday. As it turned out, lack of snow was the least of my worries.

During winter, there is only one way into Zermatt, by train. Upon arrival at the train station, we hired a horse-drawn sleigh "taxi," and instructed the driver to take us up the mountain to our chalet. We were pleased with what we found: a well-built structure with walls made of logs and a steep roof designed to shed the snow. There were ample beds, a cooking stove, and both a heating stove and a fireplace. There was no need for a refrigerator; perishable items were placed outside in the snow. Outside, we found a large supply of firewood, neatly stacked as it must be in Switzerland. The view of the town of Zermatt and the surrounding valley below was spectacular.

Bernard and I walked into town in the cold of that first evening, and returned with food supplies that were sufficient to last a couple of days. We took the small sled that we found at the chalet with us, piled our purchases on the sled and pulled the loaded sled up the road. Since we planned to end each day by skiing down the road into town, there would be plenty of opportunity to purchase whatever we needed later in the week.

During our hike back up the road, snow began to fall.

The snow continued to fall all night and throughout the next day. When the snowing finally stopped, approximately two meters had accumulated according to the information provided by the local radio station. I walked outside to explore a little and sank in deeply; the snow came up nearly to my chin. I was able to move forward, but only with the combined effort of attempting to walk aided by simultaneously pulling myself forward with my arms moving in a slow breast-stroke-like swimming motion through the light powder. Our son, David, then six years old, came out and immediately disappeared into the snow. I could hear him laughing, but I had to reach in, find him, and pull him out.

We discussed our predicament. What should we do? Our supplies were sufficient only for the remainder of that day. We learned from the radio that Zermatt was isolated from the rest of the world; the trains were not yet able to get in and out. Some supplies were being brought by helicopter. The news on the radio assured us that the first helicopters coming in would be bringing in new supplies of beer — first things first!

Because trains were not able to come in, we reasoned that even though all hotels would be fully booked for the holiday week, there would be hotel rooms available. We decided to abandon our chalet and plough our way down the road into town. With each of the boys between two adults, we went single file along the snow-covered road. Bernard and I took turns in the lead using the breast-stroke-assisted walk to make progress.

Shortly after we started on the way, we heard a loud roaring sound and looked up to see an avalanche falling down the mountain on the other side of the valley. The momentum and the quantity of moving snow built up as the avalanche progressed down the mountainside. From our distant view, we could see that the avalanche finally stopped near the edge of Zermatt and that it had completely covered the train tracks. The power of the moving snow had been sufficient to move and "un-track" a locomotive and several passenger cars standing near the station as if they were part of a child's toy train.

The journey down the road into town was actually more fun than difficult. In Switzerland, there is a restaurant every 100 meters along any such road in the vicinity of a mountain resort. We stopped for coffee and subsequently for lunch along the way.

We were able to find rooms in a small and simple hotel with half board; we had lunch every day up on the mountain. The price was within the range that we could afford. Thus, despite the change in plans and the loss of our chalet, we managed very well; our time in Zermatt was filled with learning to ski and with enjoying the mountains.

There are three different major ski areas that one can access from Zermatt by telepherique, by train or by the various lifts up to the top of the mountain. Once up into one of these areas, there are huge ski terrains, most of which are well above the tree line. Thus, on a sunny day, one sees an expanse of white snow fields under blue skies, and one sees the Matterhorn, easily identified by its unique jagged shape, surrounded by a range of spectacular mountain peaks. At the end of the day, we had a choice of either skiing all the way down into town and almost to the door of our hotel or coming down via the various means of available transportation.

Specific memories stand out in my memory of that Christmas week in Zermatt. A few days after our move into the hotel. I was up high on the Gornergrat and had just come down a long run, probably at a relatively easy level of difficulty, but I had managed to get down the entire run without falling. I was elated and had for the first time the feeling that

I was using and enjoying the mountain rather than fighting it. The following day, my son Peter and I came down a similar run together. At the bottom, his eyes were flashing with the joy and excitement of mastering the steep and thrilling downhill challenge, and he expressed to me the same feeling. These were two of our perfect days, and there were many more in the following years.

By the time we departed from Zermatt, Peter was doing moderately well on skis. David (three years younger) had been in ski school, but had not progressed very far. We had a plan to advance them to the next level. After returning from Zermatt, we arranged for the boys to live for two weeks in January at a boarding school above Montreux, in the mountains above the Lake of Geneva. They would speak only French, and they would have ski lessons every day. During this two-week period, Ruth and I traveled, first to Greece and then to Israel. On our return, we called the boys from the airport in Geneva. They were happy to hear from us and anxious to be rescued from their boarding school "prison" in the mountains, but they made us promise that we would bring our skis with us when we came for them. We skied with them during the afternoon of that same day. They had both made remarkable progress. I was instantly and permanently reduced to being the third best skier in our family. That status remained for many years until our two grandsons, Brett and Jordan, displaced me for third and fourth place. Today, I am #6, with my granddaughter, Alice, having again displaced me.

During the remainder of the winter season, we went skiing nearly every Saturday, often to Verbier, sometimes to smaller areas in France close to Geneva. We returned to Zermatt for the Easter Holiday week. This second time we had accommodations in a lovely small hotel. The food was outstanding, including the Fondue Bourguignonne that we ordered for our final dinner at the end of the vacation week. We enjoyed spring skiing under sunny skies. Life was tough — because of the bright sun and high altitude, our most serious concern was to avoid sunburn.

Late in the season in that memorable year in Switzerland, after many of the areas were already closed, we decided to go skiing one last time, in the area called Les Diablerets. Three gondola rides, one after the other in series, were required to take us from the town at the base, where we had a light lunch, to the very top, to the glacier. The weather was good, partly sunny and not very cold. After arriving at the top, we put on our skis and ventured out onto the ski run on the glacier. Judging by the three gondola rides on the way up, we knew we had a long vertical descent ahead of us and a long way to go.

In my memory, the entire trip down the mountain seems to be at night. This is of course not true; we started the descent in the early afternoon of a bright day. Moreover, my mental picture is inconsistent; in my mind, there is excellent visibility — I can see everything clearly — but nevertheless the entire trip down seems to take place surrounded by darkness. The events evidently determine the mental picture.

We started, in high spirits, along the piste that follows the glacier. Ruth and I were relatively close together, Peter and David were some distance ahead of us and skiing with Bernard. I casually remarked to Ruth that I was surprised that the slope was so gentle since we had come up to such a great height. Some time later, after traveling a considerable distance along the glacier, we saw the terrain change ahead of us, and we understood: It was no longer a gentle slope.

I recently looked up Les Diablerets on the internet. The description there is fully consistent with my memory. The internet description states that most of the pistes up on the glacier are classified as for beginning or intermediate level skiers, but a "single hair-raising black run plunges beneath the gondola cables" down toward the village.

Peter, David and Bernard waited for us above a very steep and narrow chute near the top of this "hair-raising black run." There were large rocks on either side that defined the edges of the skiable area. One could see well beyond this steep, narrow chute; the terrain below was wide and did not appear so challenging as to be beyond our abilities. I told the boys to go on through the chute, and they did so without difficulty. I then followed. Now we were standing together looking up at Ruth as she approached the chute.

Ruth lost her nerve. She was afraid that this chute was too steep and too difficult for her. One of the first things the instructors teach you in ski school is that you should never, ever take off your skis when on the mountain. Skis have long sharp edges; those edges cut into the snow and hold you onto the mountain. Instead of carefully side-stepping down with her skis on, she took them off with the intent of walking down through the most difficult area. From my position well below her, with the benefit of hindsight, I should have immediately sensed the danger, but I, too, was inexperienced, and it all happened too fast for me to intercede or give advice or instruction. She took one step before her feet went out from under her; she was on her side and sliding over the snow at an ever increasing speed. I can see her still, in the darkness of my mixed mental image, literally bouncing off the huge boulder that defined the side of the chute.

In the high mountains, and especially on ski slopes, one loses all sense of scale. Because one sees rocks, cliffs and snowfields that are miles away, distances and sizes are difficult to judge. I am nevertheless confident that after bouncing off the boulder, she fell a distance that was greater than the height of a person — at least 10 feet, perhaps more. She landed upright in snow that was both soft and sufficiently deep that it cushioned her fall. I threw off my skis and moved up the relatively flat area below the chute toward her as fast as I could go. She was frightened, but fortunately not injured.

We calmed her down. After some time and serious discussion, she decided she was all right and that she would be able to ski slowly and carefully down to the bottom. However, the snow was too deep and the slope too steep for her ability. After one turn she fell again into the snow and lost both her confidence and the ability to go on. Two skiers who had witnessed this sequence of events offered to ski down, alert the ski patrol and ask them to come to our rescue.

A short time later, two very large Swiss men arrived towing a long sled. Ruth was instructed to lie down in the sled; she was then strapped into the sled and covered from head to toe with blankets. Her skis were tucked in beside her. Because she had no broken bones, they guided the sled quickly down the mountain. I put my skis on, and the remaining four of us skied the remainder of the way down without too much difficulty, although we were certainly apprehensive about what we would find at the end of our journey.

People are used to seeing injured skiers brought down on sleds by the ski patrol. They look on with curiosity about what happened to this poor skier; they look on with a sense of well-being, thinking to themselves: "Thank God it is not me on that sled." The onlookers were, therefore, surprised when Ruth and her two ski patrol saviors arrived at the base: She told us that after waiting for a moment to be released from the sled, she stood up, took her skis, and simply walked away toward the bar. When I arrived a few minutes later, I found her there in the good *après ski* life atmosphere, looking relaxed, drinking a rather large brandy and embellishing her story to the bartender and all who would listen. She convinced her colleagues at the bar that the fast and bumpy ride down on the sled was scarier than the fall through the chute and into the deep snow.

There is a wonderful play called *Master Class* by Terrence McNally about the great soprano Maria Callas. Callas had stopped singing and was giving master classes to aspiring young women. *Master Class* was basically a one-woman play; Maria was on stage alone throughout almost the entire performance. At one point, however, a young woman joined her for a master class. The young woman sang a few notes, but was then quickly instructed to sit down. There were a few moments of silence after which the young woman asked Callas a question; the answer has remained forever in my mind and shaped my thinking. She asked Callas: "Why did you stop singing? Did you lose your voice?" Callas responded: "No — I did not lose my voice — I lost my nerve!"

One must never lose one's nerve. When skiing, losing your nerve will inevitably cause you to fall, and possibly hurt yourself, as was clearly demonstrated by Ruth's experience at Les Diablerets. More important to life in general, losing one's nerve will stop creativity and prevent the search for discovery. Losing one's nerve is "death" to the progress of the scientist or to any creative endeavor.

Despite the experience at Les Diablerets, Ruth continued to ski with us, her "boys," until she was well into her 60s. That she continued to ski for many years is a remarkable testimony to her strength of character. For Ruth, skiing was a special pleasure because of the beauty of the mountains and because it involved the entire family together. She had the pleasure of skiing many times with her family and even the very special pleasure of skiing with her grandchildren. For several years, we took our oldest grandsons, Brett and Jordan, on a ski-week vacation during the Christmas holidays without their parents. There was always laughter and a sense of joy at our good fortune to be able to enjoy the mountains together. Of course, she also enjoyed the *après ski*, the good food and wine, and the sense of well-being that was at the heart of our ski vacations. As long as the slope was not too steep or the snow was not too deep, she was a beautiful skier with excellent form. I think she finally quit because of tension with me. When on skis, I have no fear; I love to go fast and to be close to the edge ("risk-addiction?"). She tried valiantly for many years to keep up, but she finally decided the time had come to give it up.

Needless to say, the day on Les Diablerets was not a perfect day, but there were many other skiing days that did properly qualify as "perfect." There were the beautiful sunny days high above Zermatt, with spectacular views of the Matterhorn. There were days at Morgins, again blue skies and bright sunshine, where we skied while looking out at the face of the magnificent Mont Blanc.

There was a day on Ajax Mountain at Aspen during which the air literally sparkled as very, very light snow fell through the sunshine. That day on Ajax, I seemed to be able to literally float through the deep powder down the steepest of slopes. There was another spectacular day at the Aspen Highlands, with sunshine, blue sky and the Maroon Bells in the distance.

There were the days when I followed each of my grandchildren down the mountain for the first time, each time with tears of pride in

Ruth and Alan *"après ski"* at Vail in the 1980s.

my eyes. There was the day when I first took my oldest grandson, Brett, down "Forever" in the Sundown Bowl at Vail. When we stopped half way for a short rest to catch our breath, his eyes flashed with joy and excitement, just as his father's had done 25 years earlier. There was the day when his younger brother, Jordan, discovered the fun of jumping on skis and began to scour the mountain for any small bump to "get some air." I am basically a happy guy, with plenty to be happy about, but I rarely laugh. When skiing with Brett and Jordan, I always laugh!

There was the day when I took my granddaughter, Julia, through the Teacup Bowl at Vail all the way out to Blue Sky Basin. I had promised her that I would not take her on anything difficult, no Black Diamond Expert slopes, but I lied — and she loved it. There were many other spectacular days in the back bowls at Vail, including the day on December 24, 2005 at Blue Sky Basin when I followed all four of my grandchildren including Alice, the youngest one, down among the trees on an "experts only" run. There was the day at The Canyons in Utah in March 2006 when David and I skied eight Black Diamond and Double Black Diamond runs through the trees in two feet of newly fallen fresh powder; not only a perfect day but a memorable accomplishment.

I had two very special days skiing alone; one at Solitude and the other at Alta, both in the Wasatch Mountains near Salt Lake City. On each of these days, several years apart in time, everything seemed to come together enabling me to make effortless turns with beautiful form on the steepest slopes. I felt like I could do anything. I was even able to take the "bumps," the moguls, gracefully and at relatively high speed. On both those days, strangers complimented me as we rode back up the mountain on the chairlift to prepare for another run down.

On my 65th birthday, my son David arranged for us to go helicopter skiing in the wilderness in Wyoming, near Jackson Hole. Because we would be skiing in virgin snow on relatively steep slopes in remote areas, the guides gave us a brief course before departure on procedures to follow in the event of an avalanche.

Each run, the helicopter carried us up to the top of a new area with untracked snow below us. We got out crouching low and then lay flat, in deference to the whirling blades, until the helicopter took off. On the very first drop-off, I felt the ground shake just as the helicopter pulled up from the snow; the snow on the top of the ridge where the helicopter had landed moved a little as if an avalanche were about to start — but it did not. David and I got into a fair amount of trouble with our wives over that day of helicopter skiing. They were concerned after the fact — we had purposely not told them ahead of time — over the dire possibilities of skiers lost as a result of helicopter crashes and avalanches.

Actually, I lasted for only a little more than half of the day of helicopter skiing. We skied many runs through the deep powder during the morning, beautiful and exhilarating, but at the same time challenging and hard work. I was happy to stop for lunch at a predetermined area at about 1 pm. The helicopter flew in carrying soup, sandwiches, hot drinks and cookies. I was so tired from the long morning of skiing

Alan on his 65th birthday, in the wilderness near Jackson Hole.

that after lunch, I decided to stay alone at the remote spot in the wilderness where we had stopped for lunch. The helicopter crew left me with some coffee, a warm blanket, and a promise that they would periodically fly over to see how I was doing. I lay down on the snow with my head propped up on a little snow bank, covered myself with the blanket, and slept peacefully in the sunshine while David and the rest of the group went up and skied down again, several times. Later, David told me that he had reminded the pilot and the guide that the little guy with the beard sleeping down there had just received the Nobel Prize, so they had to watch out for me. Finally, they landed at my little camp to bring me out at the end of that perfect 65th birthday day.

I have had my share of perfect days skiing, far greater in number than the 14 perfect days noted by the Caliph of Cordova. And of course, before the counting is done, I must add the days of the discovery of conducting polymers, the days in Stockholm during the magical Nobel Week in the year 2000 and the specific day of the Nobel Award Ceremony and the Banquet; all perfect days.

2

The First Indication of Creativity

I showed the first indication of creativity in the physical sciences in 1940, in Akron, Iowa, when I was four years old. I am quite certain that I was four; we still lived in the "red" house, I had not yet started school, and I was not yet allowed to cross the street by myself.

The red house was not truly bright red, but the shingled siding was a deep red color; it might have been called maroon in 1940, or perhaps burgundy today. It was a simple rectangular house with a sidewalk that came directly up to the four concrete steps that rose to the front porch. The sidewalk for pedestrians that ran parallel to the street was perhaps 20 to 30 feet from the bottom step. The front porch spanned the full width of the house; the porch floor was painted grey. The roof was not flat, but symmetrical with a peak in the exact center. One entered through a door directly in the center and directly below the peak in the roof into the living room, which was separated from the dining room by a wall with a large square opening. The kitchen was in the back. There was a hallway off the dining room on the left side (when facing the house from the front) that led to two bedrooms, with a bathroom in between.

On a beautiful, sunny, warm spring morning I was playing on the sidewalk near the front steps with a set of toy soldiers that I had received as a gift, perhaps as a birthday present. I had already spent many hours playing in the house with this group of toy soldiers. This day, however, was the first time that I had taken the toy soldier army outside. The special feature of my toy soldier set was the small cannon supplied with my small army.

The cannon actually "fired": The cannon barrel was hollow, and there was a spring and lock mechanism. The "artillery shells" were nothing more than short, thin smooth sticks. Such a toy would not be permitted for sale today, for it was indeed potentially dangerous in the hands of a small child; the ejected stick could cause eye damage with resulting litigation by parents. Fortunately, however, I did not know this, and in those days parents were less aware of such potential dangers.

The game was to line up the soldiers along the sidewalk, to load the little cannon, and to fire at the soldiers. The toy soldiers were somewhat unstable so that if struck by the artillery shell, they would fall over. As in a bowling alley, if one arranged the soldiers in a tight pattern, a hit could take out the whole bunch with one firing of the cannon. I learned that the cannon could be tilted back so that I could vary the range. I quickly realized that by shooting from the steps rather than from the surface of the sidewalk, I could significantly increase the range. In reality, I was learning physics — actually I was doing experimental physics. I had a great time shooting my cannon at the toy soldiers without realizing that I was simultaneously learning physics.

However, little boys become quickly bored!

Then came the idea — I could use matches for the "artillery shells!" When a match hit the cement of the sidewalk, it would light — it would explode like a real artillery shell. The vision was clear in my mind. I immediately went into the house and to the kitchen where I found a box of matches. I took them outside and the fun began.

The very first match that I shot burst into flame as it hit the sidewalk. If I aimed correctly, the lighted match would bounce off the sidewalk and knock over one, or sometimes many, of the toy soldiers.

I was having a wonderful time and nearly got through the entire box of matches before I was caught in the act. Mother came out the front door and asked in a stern voice: "What in the world are you doing?" Of course she knew very well what I was doing, and she quickly shut down the operation. The matches were taken away, and I was sternly admonished that I should never, ever again play with matches. Although this ended abruptly, I was not seriously punished. When I think of these events today, I am certain that I remember a slight smile on her face and a shaking of the head that suggested that I had been caught, but caught doing something clever.

Perhaps the greatest pleasure of being a scientist is to have an abstract idea, then to do an experiment (more often a series of experiments is required) that demonstrates that the idea was correct; that is, that Nature actually behaves as conceived in the mind of the scientist. I have been fortunate to enjoy this experience many times in my life as a scientist. This process is the essence of creativity in science. My "experiment" with the cannon, the matches and the toy soldiers was the very first time that I experienced this intense pleasure. What I had thought should happen, actually did happen!

I have told this story many times. Ruth knows all the details and can tell it as well as I do, with even more embellishments. Many years later, Ruth and I were in London. I had an obligation to deliver a lecture on conducting polymers at a scientific conference at the Imperial College. Ruth had the afternoon free and used the time for shopping and wandering around the Mayfair district of central London. When I returned to the hotel at the end of the day, she was waiting with a big smile on her face. Without saying a word, she handed me a small box that was carefully wrapped. I opened the package to find a wonderful toy cannon. This toy cannon was somewhat more sophisticated than the one in my memory; this one was a realistic-looking miniature of a real cannon. There was a clever mechanical mechanism to raise or lower the angle of ejection of the projectile. Again, the cannon actually functioned: The cannon barrel was hollow and there was a spring and lock firing mechanism. Also in the package that she gave to me was a small box of wooden matches that she had purchased at a tobacconist's shop. The matches fit nicely into the barrel with the head of the match just sticking out. I cocked the pin, and shot one of the matches across the room. Fantastic! I could not wait to return home so that I could re-live my memory in new reality.

After returning from London and spending the necessary time to clear the waiting pile of tasks from my desk, I went out onto the back steps of our house in Santa Barbara, by the pool. I had no toy soldiers, but that was not the point. What I wanted to see was the vision of the matches exploding into fire when they hit the cement surface. But — the first one did not light, nor did the second. I got frustrated and shot directly at the wall of the step from close range. Good velocity, big bounce, but no fire — only disappointment. I returned to the house for a different box of matches. The result was the same — anticipation but no fire. The answer, as it turned out, was simple. Safety matches had been invented. The game was over and must remain, as it had for many years, in my memory.

Shooting the matches to make the game more real and more fun was the first sign of creativity and the first sign that I might have what is required to be a scientist. What I had done was truly creative: I had an idea and I used my limited knowledge of the way the world works — artillery shells explode in the movies, matches light when struck, soldiers fall when shot — to create a new reality.

Artists create. Great paintings or sculptures exist only because the artists created them. Great novels exist only because the authors wrote them. Scientific breakthroughs typically result from a combination of creativity and discovery. In science, creativity and discovery are related, but they are not the same.

Henri Becquerel discovered radioactivity. He had no preconceived idea. He simply stored some photographic film plates in a dark drawer to prevent exposure to light, and he placed some small rocks (crystals of uranium salts) on them to hold them flat in the drawer. When he subsequently developed the plates, he found an image of the shape of the rocks. He realized that this image must have been created by some kind of invisible emission from the rocks. Because of his discipline as a scientist and his curiosity, he pursued this observation, and he discovered radioactivity. Becquerel did not create radioactivity; the creativity was in his synthesis of the various observations into a whole that was the discovery.

James Watson and Francis Crick deduced the structure of DNA based on a few important experimental results obtained from the X-ray experiments of Rosalind Franklin. Their first accomplishment was to understand the importance of the problem of the structure of DNA and to have the audacity to believe that they could deduce the molecular structure of DNA. They then used simple tinkertoy-like models to try to deduce a structure that was consistent with the known facts. Of course they did not create DNA, nor did they provide the essential experimental evidence that eventually demonstrated that their proposed structure was correct. The Watson–Crick breakthough was a beautiful example of creativity. When they found the answer, they knew that they were right because they immediately understood the implications of the structure that they had deduced and thereby discovered. The final sentence of their famous paper, one page in length and published in *Nature* in 1953, tells the story: *"It has not escaped our notice that the specific pairing we have postulated immediately suggests a possible copying mechanism for the genetic material."* The Watson–Crick discovery was the detailed molecular structure of the double helix; the creativity was the realization that this structure contained the secret of life. Several years were required before the precise meaning of the double helix genome was understood. Specific sequences of the base pairs that hold the double helix together were identified as the genes. A specific local sequence within a gene was identified as the code for each of the amino acids, the components from which proteins are constructed. Proteins make the cells and the cells make the body. Watson and Crick had indeed started the discovery of the secret of life.

There are many such examples, and many of these examples resulted in Nobel Prizes. Arno Penzias and Robert Wilson made microwave amplifiers that were designed to have lower background noise than had ever been previously obtained. They pointed their low-noise amplifiers toward the night sky to test the minimum detectable noise. To their surprise, the apparent "noise" was much greater than they had anticipated. Since they understood the technology that they had created, they concluded that the apparent "noise" was not simply noise but a signal of extraterrestrial origin. The creation (invention) of better technology by Penzias and Wilson resulted in the discovery of the black body radiation that is the residual of the Big Bang that occurred at the origin of the universe, a truly marvelous discovery.

Albert Einstein created the General Theory of Relativity. Based upon deep and fundamental concepts, he created the idea of curved space–time as equivalent to, and the origin of, gravity. He developed the consequences of his ideas with specific examples and predictions that might eventually lead to experimental verification, one of which was that light would bend as it passed near a large mass.

The most basic aspect of the scientific method is that a concept can be tested and shown to be true or false. During the solar eclipse of 1920, star light that passed near the rim of the sun, observable only during a solar eclipse, actually curved around the sun; the positions of these stars in the sky appeared to shift because of this curvature. Einstein's General Theory of Relativity was demonstrated to be correct, and he became a famous international public figure, the greatest scientist of the 20th century. Einstein created the theory — and Nature actually functions as he had conceived that it would. One can only imagine the smile on that famous face when he learned of the results of these astronomical experiments.

Many years later, Alan MacDiarmid, Hideki Shirakawa and I sat together in my office in the Laboratory for Research in the Structure of Matter (LRSM) on the corner of 33rd and Walnut Streets at the University of Pennsylvania. Shirakawa had just arrived to work with us as a Visiting Scientist. We invited him to join us because of the beautiful silvery films of the polymer, polyacetylene, that he had synthesized. MacDiarmid had seen these films during a recent lecture tour to Japan. When he described the polyacetylene films to me, we agreed that we should move as quickly as possible to bring Shirakawa to the University of Pennsylvania to work with us. Hideki had arrived only a few days earlier; this was the first meeting of the three of us. We looked over the data that Hideki had acquired in great detail and had an open and lengthy discussion.

It is difficult to define the moment of discovery, and even more difficult to describe that moment. Based on the infrared absorption data that Shirakawa showed us, I suggested the critical experiment; I postulated that exposure of the pure insulating films of polyacetylene to the vapor of an electron acceptor such as iodine would cause the electrical conductivity to increase to a level approaching that of a metal. I called my post-doc, Dr. C.K. Chiang, to my office and the four of us went to the blackboard and designed a simple apparatus that would enable us to test this idea. Shirakawa and Chiang then went off to the laboratory to do the experiment. Within two days, the data that they acquired demonstrated that the electrical conductivity had indeed increased by many orders of magnitude. The "discovery" days in late 1976 must certainly be added to the list of my "perfect days."

These were exciting times, the rare times that the scientist hopes for, but rarely experiences. We were doing truly interdisciplinary science. The polyacetylene films were synthesized in MacDiarmid's lab in the chemistry building and the measurements of a broad range of physical properties that define the metallic state were carried out in my physics labs in the Laboratory for Research in the Structure of Matter (LRSM) about a block away. Graduate students and post-docs were running back and forth and the results improved on a daily basis. We experimented with new dopants (better electron acceptors than iodine) and achieved higher conductivities. At one point I said to MacDiarmid that if we were able to reach electrical conductivities equivalent to that of known inorganic metals (i.e. "real" metals), I would kiss him! He muttered in response that he was not sure that this promise was an incentive. Nevertheless, within a short time, we had polyacetylene samples that exhibited electrical conductivities equivalent to that of lead and mercury (and later even approaching that of copper) — and indeed I did kiss him!

The discovery of metallic levels of electrical conductivity in polyacetylene demonstrated that our ideas were both true and revolutionary; the field of semiconducting and metallic polymers had been created. The rest, as they say, is history — and on October 10, 2000, the telephone at our bedside rang to inform me that I would be awarded the Nobel Prize in Chemistry.

3 Miracles Actually Happen!

I was born in 1936. I lived with my parents, Peter Jacob Heeger and Alice Minkin Heeger, in Akron, Iowa. Akron was a small town, perhaps better described as a village, of approximately 1,000 people. Akron existed as a local commercial center that supported the many small farms in the surrounding area. The "downtown" area was two blocks long on a single street. Akron is situated in the northwest corner of Iowa very near the South Dakota border. The Little Sioux River, which ran near the edge of town, forms the border between Iowa and South Dakota. The largest city in the area is Sioux City, Iowa, which had a population of approximately 80,000.

My father's family lived in Sioux City, so we often made the somewhat arduous 35 mile journey by car on gravel roads on Sunday morning and returned to Akron on the same gravel roads late the same day. Although these car rides lasted about one hour, I was not bored because I spent the time watching the hilly countryside and imagining that I saw Indian warriors, cowboys and U.S. Cavalry on horseback. This area had been the heart of the Sioux Nation. The stories of the events that took place during the last half of the 19th century were still very fresh, and the images from these stories were easy to call up. The Battle of the Little Bighorn, General Custer's "Last Stand," had occurred a few hundred miles away but was part of the local lore.

In Akron, my father's grocery store was in the middle of the downtown business district. Across the street on one corner was the First National Bank of Akron with the Akron Café next door. Next to the Akron Café was Marion's Gift Shop, where I went every year before

Mother's Day to buy a present for my mother. Marion's Gift Shop had a wonderful perfumed smell that I will never forget, but I always felt out of place when I went there, just as I do today when I walk through the cosmetics section of a modern department store. On the opposite corner were the offices of the *Akron Weekly News*, published once a week, on Friday, with all the local news and with advertisements from the various merchants to attract the business of the local farmers and their families during their weekly Saturday trip into town.

There was one school in Akron, kindergarten through high school, and there was one movie theater. The name of the movie theater, the Norka, was the only innovative name for any business in town; "Norka" is Akron spelled backwards. The names of all the other various business were typical: Heeger's Market, Nelson's Feed and Supply, Dalton's Hardware, Akron Café, and on and on, but the Norka was special — it had a special name and, for me, it was a very special place.

Movies were not shown at the Norka every day, only on Friday and Saturday nights with a matinee on Saturday afternoon. I typically went to the Norka for the Saturday afternoon matinee, and I often went on Saturday evening as well. The Saturday matinees were for children; they showed cartoons and serials, the latter with a new episode every week. The Saturday evening showings were for the general audience and attracted families from town as well as from the farms in the surrounding area. The only films that I remember, however, were Tarzan adventures preceded by a short slapstick comedy starring the Three Stooges. The Three Stooges films were my favorite with Tarzan and Jane a close second. I presume that the Norka owners did not play this combination every week, but I only remember laughing at the antics of the Three Stooges and enjoying Johnny Weissmuller as Tarzan swinging through the trees, swimming under water to tip over the canoes carrying bad guys, and always just managing to save Jane.

We were not a religious family. As a child, my favorite food was ham. I was, however, not allowed to call it "ham" because my parents did not want our relatives in the extended family to know that we ate pork. My mother taught me that my favorite meat was "red round steak."

We typically went to Sioux City for the Jewish High Holy Days. Sometimes, however, Mother and I went to Omaha (Nebraska) to spend time with her family there. The Missouri River forms the border between Iowa (on the eastern side) and Nebraska (on the western side). Sioux City is on the Iowa side of the Missouri River and Omaha is on the Nebraska side approximately 90 miles south of Sioux City. Our trips from Sioux City to Omaha and back were made by Greyhound Bus.

In the early '40s, vaudeville was still alive in the American Midwest. The Saturday afternoon matinee at the Orpheum Theatre in Omaha consisted of a vaudeville show followed by a movie. I loved to go to the Orpheum, to the Theatre; seeing real people on the stage was much better than simply going to the movies.

On one of our trips to Omaha for the High Holy Days, I learned that the Three Stooges were in town and would be performing at the Orpheum Theatre on Saturday. I had to go. But, alas, I was not allowed to go — I had to be in the synagogue with my family. I reluctantly went to services, but I pouted and was miserable.

Then the miracle happened! Shortly after the service had started, three men walked down the aisle, maneuvered into the row in front of us, and sat directly in front of me. Unbelievable: The Three Stooges, Larry, Moe and Curly, had come to our synagogue! I had difficulty containing myself and sat there quietly mimicking their slapstick antics in my mind during the remainder of the service. When the service was over, I even spoke to them and wished them Happy New Year.

Throughout my life, I have had the benefit of enjoying Good Luck! Even the Nobel Prize call was the result of a combination of important discoveries and Good Luck. There are many great scientists whose scientific discoveries are of the caliber required for being awarded the Nobel Prize. But, as in most human endeavors, a little Good Luck is required as well.

Everyone in Akron knew everyone in Akron; it was indeed a village. My father was known to all as Pete Heeger and, around the downtown area, his friends and acquaintances referred to me as "Little Pete." After starting school I had free rein over the town and its surroundings. Since everyone knew my father as a local businessman, as a member of the Masonic Lodge, as President of the Chamber of Commerce, and as the head of one of only three Jewish families (all grocers by the way, and all competitors), everyone knew "Little Pete" as well.

My mother, my younger brother and I moved away from Akron in 1947, after my father died. Mother and I returned for a brief visit approximately 25 years later. By this time Ruth and I were married with two young sons. We were on our way by car from Philadelphia, where we lived and where I enjoyed the many-faceted experiences of academic life as a Professor of Physics at the University of Pennsylvania, to Aspen where I would take part in a Summer Physics Institute. Mother was living with us at the time, and she joined us on this trip. We stopped in Sioux City to see our family, and we then made the 35 mile journey to Akron: I wanted to show my sons something of my origins.

When we arrived I quickly noticed that the "Last Picture Show" had played long ago; the Norka was boarded up and closed. The town had changed because many of the small farmers had lost their land and moved on; their farms were now owned by large conglomerates. As a result, the special role of Akron as the local community for this farming region had diminished. The building that had housed Heeger's Market was gone.

Nevertheless, when we stepped out of the car in front of the vacant lot where Heeger's Market had been, two people came across the street to greet us: the editor of the *Akron Weekly News* and his wife. They had seen us drive into town, and they recognized and remembered my mother. A few weeks later, when we returned to our home in the suburbs of Philadelphia, we were delighted to find in the mail a copy of the *Akron Weekly News* with the following headline: "Heegers Return to Akron."

I was an only child until I was seven; my brother Gerald was born in Akron in 1943. In my pre-school days, my mother, father and I lived in a small house a few blocks from downtown. Our house was not in the "best" part of town. There were no street signs in Akron, but the *best* street in town was called — at least by my father and mother — "Silk Stocking Avenue." When I was six years old, my father succeeded in opening his own grocery store rather than work, as he had done for many years, as the manager of one of a chain of stores owned by a group from Sioux City. Shortly thereafter, I was proudly informed by my parents that we were moving to Silk Stocking Avenue.

What does all this description of the small town where I lived have to do with the subsequent events: my life as a scientist, the award of the Nobel Prize, and the creation of the technology of "plastic electronics?" Well, it all began in Akron.

Chapter 4 — Coming to Acceptance of My Mortality

My father's illness started on Armistice Day, November 11, 1943. Armistice Day was a national holiday in honor of the signing of the documents that ended World War I. After the end of World War II, the holiday evolved into Veterans Day, as it is known today.

Armistice Day was an important holiday in Akron, Iowa. The American Legion Club was off limits to me all year long, except on Armistice Day. The Club building was located approximately two blocks from our house and on the edge of the town park in which there was a baseball diamond and an area for children to play, with swings and a slide. The American Legion Club building was not special from the outside; inside however there were attractions that I looked forward to all year. There were real one cent slot machines, and I was allowed to gamble on them on Armistice Day. Behind the building was a "clay pigeon" shooting range. I went out to watch the Legionnaires and their guests holler "Pull!" and the disc would go flying out on a seemingly unpredictable and arbitrary trajectory. Nevertheless, the shooter often hit and shattered the flying target. There were hot dogs and hamburgers for everyone.

Relatively early in the afternoon, on November 11, 1943, Mother told me that my father was not feeling well and that they were going home. I was allowed to stay and to ride my bicycle home. Since I now rode my bike freely all over town and even out into the surrounding countryside, making the two-block trip to our house up Silk Stocking Avenue was not a problem.

When I returned home an hour or two later, Dr. Madison was there. My father was upstairs resting in bed, and Dr. Madison was talking to my mother. There were two medical doctors in Akron, Dr. Madison and Dr. Breuner. Dr. Madison was our family doctor. He was a kindly old man with white hair. He was my doctor as well. When I got sick with a childhood disease, Dr. Madison came to the house. When I was due for a check-up, Mother took me to Dr. Madison's office.

My father had experienced symptoms that were initially diagnosed as a heart attack.

After Dr. Madison went out the front door, Mother calmly explained to me that my father was not well and that he would have to go to the hospital in Sioux City for treatment. There was a small hospital in Akron (my brother was born there), but the Akron Hospital did not have the facilities needed. I was told that my father had a heart problem, but that he would be all right after a period of recovery. He would, however, have to "take it easy" in the future, and he would probably not again be able to drive. I was puzzled by the statement that he would not be able to drive. Driving did not look strenuous to me; after all, the car did all the work.

The next day he was taken by ambulance to the Methodist Hospital in Sioux City, the hospital where I was born. He remained there for several months during which time my mother began to work full time in Heeger's Market and increasingly to take over the responsibility of managing the store. My father had been able to run the store with a few helpers; he cut the meat himself, and he supervised the filling of all the orders from customers. Now, however, Mother had to hire a more senior and experienced person to serve as the butcher and to help her take care of the day-to-day responsibilities. This, of course, increased the costs of running the business. Prior to this time, Mother had only helped out, particularly on Saturdays, when the farmer's trade demanded extra effort. But she seemed to be able to handle it.

Initially, she hired a young woman to be at the house and to take care of my little brother Jerry and me. Shortly thereafter she realized, however, that Jerry needed full-time care, so she arranged that he would spend a few weeks in Omaha with her sister, my Aunt Ann, and her family.

While Jerry was away, Mother and I took the train from Akron to Sioux City several times during the week and on weekends to visit my father in the hospital. I met her at the store after school, and we walked the short distance to the train station, which was at the far end of the two block business district near the gravel road to Sioux City, to wait for the "Hiawatha." The Hiawatha was a local train; I do not know where it came from, perhaps Sioux Falls or even Minneapolis in the north, but I think that Sioux City was the end of the line. The Hiawatha carried us from Akron to Sioux City and back many times. Akron was not a regular scheduled stop; the station master had to put up a signal which indicated that there were passengers waiting. We pulled out of the train station in Akron at about 3:30 pm, went from the train station in Sioux City directly to the hospital, visited with my father, had our dinner in a diner near the hospital and then returned to Akron by train in the early evening. The highlight for me was listening to songs on the juke box in the diner; 5 cents each and you could choose whatever you wanted to hear. The Irish melody "Toora Loora Loora" was the hit of the day. I learned the melody and every word of the lyrics. It still remains one of the myriad of melodies that I "hear" in my head when I wake up in the morning.

When my brother was sent off to Omaha he was little more than a year old. He remained there for about three months. But three months is a very long time when you are only at the beginning of your second year. When his aunt brought him back to Akron, back home, he did not recognize his mother. Mother was deeply upset over this; her life was in turmoil anyway and to have "lost" her little son as well was almost too much for her to bear. Although Jerry soon became a member of our family once again, he was always thereafter her "baby."

She finally cut him loose when he went away from home after high school to go to the University of California at Berkeley. He was to take the train from Omaha to San Francisco; his bags were packed and he was ready to go. Just before they went out of the house to go to the train station, she opened his suitcase and, without his knowledge, slipped an apron into the suitcase on top of his clothes. When Jerry opened the suitcase after arriving in California, he found the apron, and he found that she had used a pair of scissors to cut the apron strings.

During the time of my father's illness, I was in second grade. I have a good record of what was going on at school because every teacher was required to prepare a short article for the "School Section" of the *Akron Weekly News* for publication every Friday. Years later, the day following the announcement of my Nobel Prize, an article about me appeared in the *Sioux City Journal*. One of my early classmates, Mary Ellen Wetzler (now Mary Ellen Olson), contacted my cousin, Maita Heeger, whose name was mentioned in the *Journal* article. After obtaining my address, Mary Ellen sent me a copy of the scrapbook that her mother had prepared during the years we were classmates in kindergarten, first and second grades. What a treasure trove! The class articles from the *Akron Weekly* were carefully preserved giving me a record of my early childhood that I had completely lost. Included was a photo of Mary Ellen and me taken when we were in kindergarten all dressed up as the Royal Attendants to the Homecoming King and Queen of the Akron K-12 school.

I have many examples similar to the following "news" article which appeared in the *Akron Weekly News* in early 1944:

Second Grade

Alan Heeger spent last Thursday in Sioux City. Vincent Mc Mahon was absent on Wednesday due to illness.

The second graders are learning to spell the days of the week and their abbreviations. The "three's" multiplication tables are being learned in Arithmetic.

Gerald Erickson brought a turtle to school which the children have enjoyed watching.

The following children had perfect Spelling lessons this week: Sylvia Adams, Norma Jean Bellwood, Lozetta Bohman, Maty Ellen Wetzler, Donald Kapfer, Alan Heeger, Vincent McMahon, John Pannkuk and Donald Peters.

My spelling record was not always so perfect. Another such article states that Mary Ellen Wetzler got a perfect score, but Alan Heeger missed two! Mary Ellen always got a perfect score in everything; that is why her mother kept the scrapbook.

I was a good student and typically got A's on my report card except occasionally when I had to be satisfied with less than a perfect record in "Deportment." Mother always had to sign the report card, and she had the opportunity to write a short comment if she chose to do so, I suppose to assure the teacher that she had actually seen the report card. I was instructed to return the signed card to my teacher. Mother always wrote the same comment: "Alan will try harder next term." This comment always annoyed me. I had received all A's — what better could be expected? From the beginning I was confronted directly with the requirement of meeting rigorously high standards!

Although my father was in the hospital, my life went on more or less as before with the exception of the regular trips by train to Sioux City and back during the week and on Sundays. Saturdays, we stayed in Akron because of the busy day at Heeger's Market.

Eventually, my father returned home to Akron, but he was not well and spent most of his time in his bedroom upstairs. Over the following months, he visibly deteriorated and eventually was no longer able to make the trip up and down the stairs. At this point, Mother re-made the dining room into a bedroom for him, and had a toilet and small sink installed in what had previously been a closet adjacent to the dining room.

There was one especially good feature about this new arrangement. The new little bathroom had no windows. When I went in and closed the door without turning on the light, it was pitch black inside. This was the perfect place for me to play. Again I had received a gift, again probably a birthday gift. This time, however, the gift was intended to be an introduction to real science. In the box were a number of rocks in various appropriately labeled small drawers and a "black light"; that is, an ultraviolet emitting light bulb. I spent many hours in that little bathroom when I was eight years old. First, I stood on the toilet seat and reached up to the light fixture that was mounted on the wall. I unscrewed the regular light bulb and replaced it with the "black light" bulb. Then I closed the door, turned on the ultra-violet light and carefully examined each of the rocks in turn. They were marvelous; they glowed in the dark with a variety of beautiful colors. This was my first introduction to luminescence and phosphorescence, both of which played a role in my life as a scientist many years later. These luminescence "experiments" truly fascinated me. I went into the dark room many times and examined the glowing rocks in great detail while wondering why light with such beautiful colors was emitted. I would like to think that this was in indication of my future in science, but more likely it was simply childhood curiosity.

I learned that my father was going to die when Mother told me so while supervising me one evening as I was getting ready to take a bath. We were in the bathroom and the door was closed. She began to cry and said to me: "We are going to lose our Daddy." Those were the precise words. I did not argue with her or question her, but I wanted to do so because I did not yet believe that he would die.

According to what I learned later, my father died of congestive heart failure. Toward the end of his life, he was unable to get out of bed, and he had lost a great deal of weight. He died on June 26, 1945, one day after his 45th birthday. At the end, Mother was at his bedside and Dr. Madison was there. I had been sent out onto the front porch, but I knew what was going on. After he died, Dr. Madison brought me in, and I found my mother sobbing over his wasted body.

In the short biography that I wrote for the book published by the Nobel Foundation in 2000, I stated that my father died the same day that Franklin Roosevelt died. I subsequently learned that President Roosevelt died two months earlier, on April 12, 1945, but I have always associated the two deaths as being on the same day. Perhaps it was a way for me to increase in my mind the importance of my father's death.

The funeral was in the Conservative Congregation synagogue in Sioux City. At that time, approximately 600 Jewish families lived in Sioux

City, Iowa. It seemed to me that everyone was there. Peter Heeger had grown up in Sioux City, and he was well loved. And of course there was a contingent of people from Akron in attendance, as well as members of my mother's family from Omaha.

I remember every aspect of the funeral with great clarity, but this clear memory only returned to me on June 26, 1995, 50 years after the actual date on which my father died. In June 1995, I was involved in a hundred things. My science was going well, I was living in Santa Barbara and teaching at the University of California at Santa Barbara, where I had a large and active research group, I had started a company with the goal of commercializing our discoveries in the area of conducting polymers. Ruth and I had just finished completely remodeling our house. I had not been thinking about my father's death nor was I consciously aware that the 50th anniversary of his death and his funeral were approaching. Nevertheless, when I awoke on June 26, 1995, I was there, in the synagogue, in Sioux City, and I remembered it all with absolute clarity.

I remembered sitting with my mother in the front row directly in front of the closed casket. I remembered my grandmother wailing and pleading to God to take her instead. I remembered the Rabbi's eulogy; he compared my father's life to a candle burning at both ends; such a candle yielded a great deal of light but burned for a relatively short period of time.

I remembered that at the end of the funeral service, the casket was wheeled out to the front entrance of the synagogue and opened before being carried out by the pallbearers to the waiting hearse and transported to the cemetery. There were many people crowding around the open casket, but space was made available and someone brought me to the front. I was lifted up to the open casket, and someone — whoever was holding me — told me to say good-bye to my father. I was confused and did not know what to do. I leaned deeply into the casket, and I kissed my father goodbye.

Even though I was only nine years old when he died, I have a few specific memories of him. The earliest was a bathtub scene. I was very young, perhaps only three years old. He and I were taking a bath together; I was facing him in the bathtub. At his instruction, we both got up on our knees and he taught me how to wash my "tush." The only reason this somewhat intimate family scene is etched into my mind is because we were all laughing; my father was laughing, and I was laughing. My mother was in the bathroom as well and laughing so hard that she could hardly breathe.

He taught me the value of money. He sent me off to the barbershop one day to get a haircut with a 50 cent piece in my hand. I jauntily tossed the coin up in the air and caught it again as I walked toward the barbershop. Just in front of the drugstore I missed the catch; the coin fell through my fingers and went through a grate in the sidewalk. Never mind, I went back to ask my father for another 50 cents. This was the only time that I remember him being angry with me. He insisted that I show him what happened and where I had lost the coin. When we got to the grate in front of the drugstore, we peered down and there at the bottom of the pit was the silver coin. My dad then simply lifted the grate and told me to climb down and retrieve the 50 cents. Although it was dirty and filled with a lot of yucky junk in there, I did bring out the 50 cent piece. Needless to say, I never forgot the incident.

He liked to have a drink; I take after him in that respect. When we went to visit family in Sioux City, we sometimes stopped on the way out of town before we got back onto the gravel road to Akron. Mother and I would wait in the car for a few minutes while he went into the local bar for a "short snort."

Whenever he went out of the house in the evening, I would complain as little boys always do and ask him where he was going. The answer was always the same. He was "going out to see a man about a horse." I never questioned that response.

The busiest business day of the week at Heeger's Market was Saturday. On Saturday morning and early afternoon, the customers from the nearby farms would come into the store to chat for a few minutes, and they would leave with my father the list of grocery, meat and produce items that they needed. By the end of the day, the entire floor area was covered with the completed orders carefully arranged and marked with the family name of the customer. Heeger's Market stayed open late on Saturday evening so that the farm families could enjoy the evening in town. As a result, I often went to the movies on Saturday evening, sometimes had a hamburger — on white bread with yellow mustard — and a coke at the Akron Café across the street after the movie, and then came to the store and waited with my parents until the last of the orders was claimed.

After closing the store, my parents sometimes took me with them to a bar around the corner. I was allowed to literally sit on a barstool at the bar while they relaxed and chatted with friends and with the bartender. When I looked straight ahead from my barstool, I saw a large mirror. In the mirror was the memorable scene of "Custer's Last Stand"; a painted mural that covered the entire wall behind me. The scene showed dead U.S. Cavalry soldiers all over the ground. Their horses were mostly down, but a few were struggling to rise. There were dead Indians all around as well. In the center, however, was General George Custer alive and bare headed with his blonde curls flowing in the wind. He had his sword in his right hand and his pistol in his left hand. He was out of ammunition. The Indians were attacking and coming at him from all directions. Every detail of Custer's Last Stand is inscribed in my memory.

On Saturday, or during the summer, I often worked at Heeger's Market. My "job" was actually quite interesting. Farmers would bring in eggs laid by their chickens and either sell them to my father or use them as barter. Before they could be resold to our customers by Heeger's Market, they had to be checked; rotten eggs or fertilized eggs containing unborn chicks had to be sorted out and thrown away. This was accomplished by "candle-ing" the eggs. In a trailer behind the store, a "light box" had been constructed. The light box was a fully enclosed wooden box mounted onto the wall with a lighted bulb inside. To be precise, the box was not fully enclosed but had a small hole cut into its front wall.

The diameter of the hole was slightly smaller than the diameter of an egg, and the hole had been carefully smoothed with sandpaper. When I placed an egg over the hole, I could see inside the egg. If the egg was all right, it appeared uniform and clear. The bad ones, those that were rotten or contained an unborn chick, were obvious under this close and elegant scrutiny. This job was actually fun. Perhaps I experienced it as a kind of scientific experiment. Moreover, "candle-ing" eggs had a reward. When I found a bad one, I would go outside and throw the egg across the vacant lot next door to smash against the wall of the neighboring building. Sometimes, if I really got bored and there were no "bad ones," I would toss one against the wall anyway!

The months following my father's death were difficult for me and of course difficult for Mother. I had bad dreams. She cried a lot. In the evening, she often sat in front of the mirror of her dressing table and sobbed out loud saying over and over again that she wished that she were dead. I was the only one there to console her, and I tried to do so, but I was inadequate to the task.

My mother and father had been married for 16 years. To me at that time, 16 years of marriage was a lifetime, but they loved each other deeply and had looked forward — as we all do — to a long life together. In addition, she was at a loss as to how to proceed with her life and how to find the resources to raise her two sons.

I came to grips with my mortality at age nine. I understood and I accepted the fact that I would die. I also made the first important steps toward adulthood in the period following my father's death. I took on a new level of perceived responsibility within the family, and Mother began to treat me differently.

She explained to me that without my father in the store, the business was not profitable. We discussed together the problems of remaining in Akron. Approximately a year after my father died, we made the decision to sell Heeger's Market and to move to Omaha and join her family there. During the following months, she made the necessary arrangements to sell the business and to sell the house on Silk Stocking Avenue.

We planned to make the move away from Akron just after the end of the school year in 1947, just after the completion of my sixth-grade year. My sixth-grade year was therefore special. I knew that I would be leaving and that my life was going to change. To me, Omaha was the "Big City," and I was naturally uncertain about this unknown future.

Fortunately, I was distracted from these fears and uncertainties during the sixth-grade year; I fell in love for the first time. Sylvia Adams had been in my class for several years, perhaps even from kindergarten onward. She was a beautiful little girl — I have a picture of her in my wonderful scrapbook that was sent to me by Mary Ellen Wetzler — but I had not paid much attention to her. That changed in the sixth grade. I was now eleven; perhaps some hormones had started to flow. I looked for ways and excuses to walk home with her, but I had to be careful, because one could easily be noticed by the other boys in the class and that would cause no end of teasing. I watched her during class and thought about her all the time.

The year ended with a class outing. The teacher arranged to take the class for a day in the Regional Park near Sioux City. It was a terrific day in early June; the weather was fine, we were free to roam. The children played on all the swings and slides in the park, and we had a picnic lunch. Sylvia and I however, spent much of the day going down the tunnel slide, again and again, with my arm around her waist. I think she loved me too that day.

During the trip back to Akron, I was well aware that this specific day marked the end of the first period of my life. My childhood was over, but I was in reality and in many ways still a child.

Mother received very little for Heeger's Market; she basically walked away from it with the agreement that the new owner would take on the debt obligations. She sold the house on Silk Stocking Avenue for $2,500. After the move to Omaha and an initial period of a few months in a rented apartment, she used the $2,500 to buy a two-bedroom house on Lincoln Boulevard. We lived in the Lincoln Boulevard house through my final two years of grade school and through my three years of high school. The Lincoln Boulevard house was not a large house; we filled it up. In addition to Mother, Jerry, and me, Mother's sister Ann Soskin and her daughter Harriet lived with us there. Harriet's brother Milton Soskin stayed there as well for long periods of time. Finally, Mother's brother Lou Minkin stayed with us for two years after his divorce.

Mother worked as a checker in a local "Hinky Dinky" supermarket, and Ann worked at a ladies apparel store in downtown Omaha. These were not highly paid jobs, but we had enough money; there was always plenty of food on the table, and we all dressed well. Overall, I remember a lot of laughter in the Lincoln Boulevard house — usually caused by something Jerry did or said — and no sense of stress.

When we moved from Akron, the only thing I had that belonged to my father was a ring with a small diamond mounted in a somewhat old-fashioned setting that he always wore. Although I was not allowed to wear it yet, Mother had saved the ring for me and told me that his ring would be mine to wear after my 13th birthday; that is after I "became a man" on the occasion of my Bar Mitzvah. In fact, at my insistence, I began to wear the ring on my 11th birthday, shortly after we moved to Omaha. Although nearly 70 years have passed since he died, I continue to wear that same ring on my right hand today and when — on occasion — I notice it on my finger, I think about my father.

Because I was only nine years old when he died, my father played little or no role in my education or in my choice of career. Mother naturally took on this role and guided me to where I am today. In my earliest memories, she lectured me on the importance of education, and she insisted that I would "go to college," but she had no concept or suggestion that I might pursue a scientific career. She knew that I was good at math (although I was still only doing sixth-grade arithmetic), so she suggested that I consider a career as an actuary. I really did not understand what that meant, but when someone asked me what I wanted to be when I grew up, I proudly answered that I was going to be an actuary. My later evolution into science was completely self-motivated and did not occur until I was in my second year at the University of Nebraska.

She was proud when I completed my Bachelor of Science degree, and even more proud when I completed the requirements for the PhD. She watched me grow as a scientist. I told her about our early discoveries in the area of organic conductors, and that these discoveries were important in that they involved new materials with wholly new properties. I told her that these research results were sufficiently new and original that there might be a chance, one day, for me to be awarded the Nobel Prize. She died in 1993 at age 85, well before the award of the Nobel Prize — but she believed in me and would not have been surprised.

5 The Wart and the Penny: The Importance of Uncorrelated Events

Behind the red house in Akron — at the back end of our huge yard — was an alley. The alley ran parallel to our street. Our street was relatively near the edge of town, so the next street farther away from the downtown area had few houses, but each was on a big lot. These houses on the other side of the alley were small, simple houses and not as nice as our house. The large lots were really something like small farms. My alley ran behind these little farms. There were vegetable gardens, chickens and goats running loose, and one of the lots had a pig pen.

As a small child I liked to wander down the alley to see the animals and play and imagine that I was on some big adventure. Typically, however, when I got to the alley at the end of my backyard, I turned right. There was a reason that I turned right. If I turned left I would go toward the house where the Old Lady lived. To me, she was really old, she had teeth missing, and she had deep wrinkles on her face. She always wore a black dress. I had never actually spoken to her. I was terrified of her and always did my best to go the other way to avoid going near her house.

I recently read — again after many years — Harper Lee's wonderful novel, *To Kill a Mockingbird*. Harper Lee was able to convey the terror that Jem and Scout had of going too close to the Radley House. They wanted Boo Radley to come out, but they were terrified that he actually might come out. My fear of the Old Lady in the black dress was something like that expressed by Harper Lee, but my fear was simpler: I did not want her to come out, or if she did, I did not want to be nearby.

During the summer of my fifth year, a small wart appeared on the second finger of my left hand; not the "pointer" finger but the one next to it, the "big" finger. I do not know how I got the wart, it just appeared. If I opened my hand, the wart was right in the center of the center (second) section of my big finger. As a result, especially since I am left-handed, the wart was easily irritated and difficult to keep clean.

I must have complained to my mother about the wart, as children do about such things. Alternatively, she might simply have seen it, and seen that it was easily irritated. Mother told me that I should get rid of the wart, and she told me how to accomplish this cure. She gave me a shiny penny, and she instructed me to go down the alley to the Old Lady's yard. I am certain that she referred to the Old Lady by name, but I do not know her name. I will refer to her as the "Old Lady" because that is how I remember her. I was to give the shiny penny to the Old Lady, I was to ask her to rub the shiny penny on my wart, and by doing so, the Old Lady would make the wart go away. Under no circumstances was I to keep the penny. I was specifically told that after the Old Lady rubbed the wart with the penny, she was to keep the penny.

When I think of this today, I find it difficult to understand why Mother set these events in motion. She was a modern, intelligent woman. After completing high school in Omaha where she grew up, she had earned a scholarship to continue her education at the University of Nebraska. She was forced to turn down this opportunity to go to university because she had to take a job to help support her family. Her father (my grandfather) had died before she finished high school, she had three younger siblings, and her mother (my grandmother) did not have any other source of income. Mother always regretted that she had not been able to go to university, and she had great respect for education.

Thus, my mother was not a simple country girl, but rather an intelligent, educated woman. She was not particularly superstitious — although she told me never to walk under a ladder. Therefore, I cannot imagine that she actually believed that the rubbing of the shiny penny on my wart by the Old Lady would make the wart go away. I can only surmise that she wanted me to confront my fear and see for myself that the Old Lady was really a nice woman who liked to talk to little boys.

Whatever her reasons, I followed her instructions as a good boy and a good son should always do.

The event was quite simple. I turned left when I got to the alley, and I went down the alley to the Old Lady's yard. I showed the wart to the Old Lady, and I told her that my mother had sent me to her so that she could make the wart go away. I told her that she was to keep the penny. There must have been a conversation, but I do not remember anything that she said to me. She accepted this request, seemingly without surprise, and she did exactly as I had asked. It hurt a little, but not much. The rubbing was not so vigorous that I cried or expressed pain; the skin on my finger did not break.

The wart went away. I do not remember whether it was a few days later, or a week later. But, after a relatively short time, the wart was gone.

I recall being somehow doubtful even then that the rubbing with the penny had caused the wart to go away. But facts are facts — and the troublesome wart was gone. Even today, the big finger on my left hand is perfect.

This should have been an important lesson in my life. If I had been a little smarter or a little more mature, I would have realized that although the rubbing of the penny on my wart and the subsequent disappearance of the wart both truly happened, the implied cause and effect relationship was non-existent. The rubbing of the penny and the disappearance of the wart were not related.

This kind of thing happens over and over during the course of one's life, and a scientist, in particular, must carefully watch out for such apparent, but not genuine, cause and effect relationships. If the scientist does not recognize such uncorrelated events, serious trouble can follow. I learned this lesson the hard way more than 30 years later.

In the early 1970s, I had started to do research on organic conductors. A graduate student in the chemistry department at Penn, Anthony (Tony) Garito, had brought to my attention the TCNQ molecule (the full chemical name is not important here) and the early work that had been done on the TCNQ compounds at the DuPont Central Research

Laboratories. TCNQ molecules are flat and planar; in crystals of TCNQ compounds, the molecules typically stack upon one another like columns of poker chips. There were hints from the early work that some of the TCNQ compounds were relatively good conductors of electricity along the TCNQ stacks, but highly resistive to carrying electricity perpendicular to the molecular stacks. Hence, in terms of their electrical properties, these compounds were approximately one-dimensional. It seemed to Garito and to me that the TCNQ salts might offer a special scientific opportunity; a new direction for solid state physics that was previously unexplored. The quasi-one-dimensional nature of the electron transport along the molecular stacks was unusual. Perhaps, we could even test some of the rather esoteric ideas in the literature of theoretical physics that were specific to "one-dimensional metals." I invited Tony Garito to join my research group in a post-doctoral research position after he completed the requirements for his PhD.

We had many difficulties in starting up this project. We had to learn how to synthesize the molecules, and we had to learn how to grow the crystals. We were frustrated over the fact that the needle-like crystals were very small, a few millimeters long at most and typically a fraction of a millimeter across. Although I had enlisted two graduate students, Larry Coleman and Marshall Cohen, into my research group to carry out the basic electrical measurements as a function of temperature on these new organic compounds, progress was slow because of the difficulties of attaching wires to the tiny delicate crystals using silver paste (a conducting glue). With practice and effort, they did learn to mount the delicate crystals and to attach the requisite four wires, although there were occasions when Marshall Cohen successfully mounted a whisker from his black beard rather than a crystal of the TCNQ compound.

Our first publication in this area was on a particular compound known as NMP-TCNQ. We were able to establish that this organic material was indeed a quasi-one-dimensional conductor and that its electronic structure was close to the boundary between metallic and insulating behavior. Metallic behavior had never before been observed in an organic solid (one comprised of molecules that are based on the chemistry of carbon). Moreover, the transition from metal-to-insulator was a topic of keen interest in the solid state physics community during the early 1970s. As a result, these initial studies immediately had impact. We were onto something new and exciting.

The uncorrelated events occurred during our subsequent studies of a related, but different compound, TTF-TCNQ. Remarkable and previously unheard of for a solid made up of only organic molecules, TTF-TCNQ seemed to have all the properties of a true metal. Moreover, the electron mobility was excellent along the molecular stacks, but orders of magnitude lower for transport of charge perpendicular to the molecular stacks. Thus, TTF-TCNQ was the first real physical system with electrical properties that approached those of an idealized one-dimensional metal.

Theoreticians had predicted that one-dimensional metals would be unstable. Rudolf Peierls, a German theoretical physicist who emigrated to England in the Nazi era before World War II, had published a paper showing that a structural distortion was expected to occur in a one-dimensional metal and that the predicted variation in the spacing between successive molecules would necessarily convert the metal into an insulator. The demonstration of the reality of the "Peierls transition" in a real crystal would be an important scientific contribution.

Motivated by dreams of the Peierls transition, we (Coleman, Cohen and Garito with me as the "slave driver" and cheerleader) carried out measurements of the electrical conductivity of TTF-TCNQ as a function of the temperature. The data began to emerge with clear evidence that there was indeed a metal-to-insulator transition at temperatures slightly higher than that at which nitrogen became a liquid. As a TTF-TCNQ crystal was cooled below room temperature, the electrical conductivity first increased as expected for a metal. At lower temperatures, the conductivity went through a sharp maximum and then dropped precipitously toward zero. It was the sharp maximum that got me.

We found that the data were reproducible over most of the temperature range studied. However, right near the transition, the height of the sharp maximum varied significantly from crystal to crystal. Since the crystals were known to be very delicate, I suggested that the variation might arise from stress on the crystals by the wires that were attached to them. The test that I suggested would be to use much finer wires and specifically to use very fine wires made of gold because pure gold is very soft and malleable. The fine soft gold wires would relieve the stress, I reasoned, and allow us to observe the intrinsic behavior.

This turned out to be a classic example of the wart and penny effect. The first experiments carried out after implementing the suggested changes in experimental procedure yielded remarkable results. The sharp peak in the electrical conductivity was higher than we had ever seen before. It even seemed to diverge toward infinity before crashing down toward zero in the lower temperature insulating regime. We suggested that the extraordinary behavior and the apparent divergence might be a signature of superconductivity at high temperatures. John Bardeen, the only person to be awarded two Nobel Prizes in Physics, the first for the discovery of the transistor and the second for creating (with his colleagues Leon Cooper and Robert Schrieffer) the BCS Theory of Superconductivity, suggested to us that what we had found might arise from coherent "sliding mode conductivity"; a collective mechanism that had been proposed years before as a possible explanation for the remarkable phenomena associated with superconductivity. The sliding mode concept was not the correct description of superconductivity; the BCS theory (published in 1957) contained the correct answer. Nevertheless, as a new kind of collective transport, the sliding mode mechanism was fundamentally interesting in its own right.

The problem, however, was that the use of the gold wires and the subsequent observation of the giant peak in electrical conductivity were not related. We were fooled by uncorrelated events; I had taken too seriously my belief in the cause–effect relationship.

A serious and major scientific controversy grew around the question of the reality of the peak in the electrical conductivity in TTF-TCNQ: How large was the intrinsic peak in conductivity and what was the underlying physical mechanism for the maximum in conductivity that occurred right near the metal-to-insulator transition?

This was heavy stuff; people shouted at each other at scientific conferences. For example, during a conference at Lake Arrowhead in the mountains near Los Angeles in late 1973, one of our "rivals," Professor Aaron Bloch from Johns Hopkins University made a wisecrack directed at me during his lecture. It was in the evening after a long and difficult day. I jumped from my seat in the first row, furious about his comment, and immediately started chasing Bloch around the lecture table! Aaron was tall with very long legs, and he managed to avoid my attack until I had settled down and eventually returned to my seat. This was indeed fortunate, because I had no idea of what I would do if I actually caught him!

The highest conductivity peak values that we had measured were soon identified to be artifacts. With great effort, we and many others in the scientific community got the facts right. The peak was real but finite, and it did result from the fundamentally new physics of sliding mode transport near the Peierls transition. Eventually other researchers found similar phenomena in different compounds with similar quasi-one-dimensional structures.

This was a very stressful period for me. Powerful scientists were hounding me and loudly attempting to publicly discredit me at scientific conferences and in the scientific literature. At one point, a paper was published in the prestigious scientific journal, the *Physical Review*, with no fewer than 30 authors who stated, in essence, that we were wrong. I thought my career as a scientist was over. I was at a seriously low point, mentally "down," but — *I did not lose my nerve.* I insisted that we find a new method to measure the electrical conductivity, especially a method that did not involve electrical contacts with wires. We proceeded to develop the use of microwaves (the same wavelength range as used in radar) to measure the electrical conductivity as a function of temperature, and we were able to obtain convincing and reproducible results.

The only way that I can describe my despair during this period is through my attempts to express my feelings with poetry, as written at that time. I have only attempted to write simple poems three times in my life. The first time was when I was seven years old and in second grade. The resulting poem turned out to be my very first publication: It was published in the *Akron Weekly* newspaper. The second time was when Ruth and I and our sons first visited Grindelwald in Switzerland. The staggeringly beautiful views of the Jungfrau caused me to try to express my feelings in words. Fortunately, perhaps, I did not keep the Grindelwald poems. The third and last time I tried to express my feelings through poetry was one evening in late 1973.

Science, Life and People

I do not understand
Many things
Of Science, Life and People.
Enormous efforts expended to no end, it is easy to be cynical.

Science is truth — and argument
Ideas are exciting — and make enemies.
Creativity requires taking chances.
Sometimes one loses.
Sometimes one wins
   and no one even cares.
That is worse.

But it is what I know
   and it is what I do.
As long as it keeps me up at night
I will pursue.

I include this poem not because I think it represents great literature, but because the poem expresses a little of what I was feeling at that time.

Ruth's character and insight once again came through. She understood what I was going through and understood the importance of going forward and resolving these scientific issues. So she found a way to help me through the dark tunnel.

Dr. David Sachs was a friend and our neighbor in the suburbs of Philadelphia where we lived. His son, Paul Sachs, and my son, David, were close friends all through junior high and high school and remain in contact even today. David Sachs is a Freudian psychiatrist who practiced psychoanalysis. Ruth invited Dr. Sachs, his wife Romaine and their children to our house for a weekend barbeque. This was not an accident; Ruth planned the barbeque with the specific intent of getting me together with David Sachs in an informal setting. David had seen me a number of times in recent weeks; he knew that I was not in good shape. He and I talked a little, but in reality nothing much happened. He simply asked if there were something he could do to help me. I thought for a moment. I then thanked him for caring enough to try to help, and I responded that I thought I would be all right. This was certainly David Sachs's fastest successful psychoanalysis. That moment was somehow a turning point for me. The very fact that someone cared was sufficient to give me strength and to get me going again. The drive, the leadership, the curiosity, the self-confidence and the creativity returned — and I went on with my life.

Eventually with hard work, some good fortune and help from and collaboration with an outstanding French physicist, Robert Comes, we demonstrated without ambiguity that we had discovered the Peierls transition and the associated soft-mode structural phase transition. This is an example, repeated many times subsequently in my scientific life, where interdisciplinary collaboration saves the day. At that time, Comes was the only person in the world with expertise in an area of X-ray scattering that would enable us to observe the structural distortion that was the signature of the Peierls transition. His experiments demonstrated that the wavelength of the new periodicity within the organic TTF-TCNQ crystal (observed in the insulating phase at low temperatures) was incommensurate with the original crystal lattice. This discovery provided fundamental insight into the charge distribution along the TTF and TCNQ stacks, and it provided a wholly new methodology for determining the charge distribution in this remarkable new class of quasi-one-dimensional organic compounds.

In retrospect, we had taken a risk, and we had made important scientific progress. We had stimulated many other scientists to enter into a new field within condensed matter physics on the boundary between physics and chemistry. Important and exciting new ideas were generated, such as the demonstration of "sliding mode conductivity." The overall effect was overwhelmingly positive and attracted the interest and participation of scientists in Europe and Asia as well as in the United States. I was the recognized leader. I had personally learned an enormous amount of physics and chemistry in the process. Our work was flawed, however, by the penny–wart effect. We could have accomplished what we did accomplish, and perhaps even more and in a shorter time, had we been more alert to the possibility of unrelated events.

On the other hand, it is quite clear that the risk and the resulting controversy had positive impact. I matured as a scientist and began the journey into interdisciplinary science that would define my entire career. Although being at the center of that storm was unpleasant and even disturbing for me, the "storm" itself focused attention on a new area of science and stimulated scientific progress. The importance of the science that evolved from our TTF-TCNQ studies was subsequently recognized by the award to me of the Oliver E. Buckley Prize in Solid State Physics by the American Physical Society.

The life of a scientist is interesting, fulfilling and deeply rewarding. However, even great scientists seldom become wealthy and very few receive the highest honors exemplified by the Nobel Prize. The fulfillment and the rewards come primarily from the personal satisfaction that results from one's scientific contributions and accomplishments, and from the recognition of those accomplishments by one's scientific peers. Recognition by one's peers, however, implies an unstated competition for scientific credit: Who did what first and was it right?

Competition for credit is ubiquitous in all creative endeavors. The "who did it first and who was right" concept does not apply directly to artists. For artists, there is no right and wrong. For example, architects want to create structures that inspire people, and the really great architects succeed in doing exactly that. However, they certainly want credit for having done so and for being the first to have done so. At the suggestion of my son, David, I walked all the way around the National Gallery in Washington, D.C. designed by I.M. Pei. When the narrow, sharp corner edge of the building came into view, I was amazed and indeed inspired that someone could imagine and build such a structure. Similarly, the structures created by Frank Gehry, the Guggenheim Museum in Bilbao, Spain, or the Walt Disney Concert Hall in Los Angeles are wholly new and original. There will be copies or at least buildings built by others that are influenced by Gehry's originality, but he is without doubt the first and has received credit within his profession for being the first to build such structures.

The issue of scientific credit played a major role in the controversies of the early 1970s that resulted from our work on the metal–insulator transition and the Peierls transition. My mother was living with us through much of this period. She was well aware of what was going on, and she was somewhat surprised by the level of competition among scientists. Competition for credit is, however, fundamental to creative people. The competition for credit among creative people is perhaps even stronger than the competition in the business world, where financial gain is the final arbiter.

At some point in the mid-1970s, she gave me a sheet of paper with the following quotation from Rudyard Kipling typed upon it:

I once knew an unimportant man who lived in a village near Kilamanjaro, where he sewed uniforms for the officers. He was a splendid tailor but he did not want to settle for that, so he spent his other hours pushing back the brush and the wait-a-minute thorns — so called because when one snared your jacket you had to stop and pull it out — searching for precious stones. When he finally found some stones one day, he fell to his knees and wept, because for the remainder of his life he would be rich and celebrated. But quickly the lions came and attacked his camp, thieves came and stole his gems, and enemies came to dispute his claims. Before a very short time had

*passed, his heart, as people say, broke. He died and was buried in
the coffin of a tailor.*

*I have no moral, only an observation: in the country of the
unexplored diligence can find the treasure, but only power can
keep it.*

In the world of competitive creative people, one must have the intellectual strength and scientific power to fight the "lions," the "thieves" and the "enemies who come to dispute one's claims." I keep the Kipling quote on my wall as a reminder — Kipling had it right.

During this stressful period, I did not lose my nerve; I continued to do experiments involving new experimental techniques. One must never lose one's nerve. When involved in any creative endeavor, losing one's nerve will limit (or perhaps even stop) the creative process. The "wart and the penny" and the experiences of the TTF-TCNQ era had strengthened my character. I would never lose my nerve in the pursuit of science.

6 "Whither Thou Goest, I Will Go"

Chapter

Ruth is a biblical name. As told in the bible, Ruth was the daughter-in-law of Naomi and the great-grandmother of King David. Her often quoted phrase is a statement of loyalty and personal commitment.

I first met my Ruth, Ruthann Chudacoff, when I was 12 years old. Ruth was a friend of my cousin Harriet; they took lessons together every week after school with the goal of achieving graceful bearing and good posture. Harriet and Ruthie (as she was called then and as I still call her today) and a few other girls their age walked around our living room in the Lincoln Boulevard house with books balanced on their heads, sup-posedly to give them the poise and good bearing expected for proper young women. Their parents actually paid real money for these lessons! I did not pay attention to these graceful efforts or to Ruthie while she was balancing books on her head.

I became seriously interested in Ruthie during the early part of our first year at Omaha Central High School; I was 14. Ruthie was beautiful, she had a great personality, she was well "developed," and she was very popular with the older boys, especially the juniors and seniors. Many guys were asking her out, and I had the impression that she went out on dates regularly. All of these guys had access to cars; some even actually had their own cars, and could take her out in style. I had no car; we did not own one. Actually, that was irrelevant because I did not know how to drive, and I was too young even to have a driver's license. Nevertheless, this was to me a disaster. I was simply and literally outclassed.

I had a Saturday afternoon job working in the "Hinky Dinky" super-market in Council Bluffs, Iowa, across the Missouri River from Omaha.

I went to work by streetcar that carried me to downtown Omaha. I then transferred onto a bus that took me across the Missouri, to the Iowa side and my destination in downtown Council Bluffs. I returned home at the end of the day by the same route in reverse order. My uncle, Lou Minkin, was the manager of the store. There was a chain of "Hinky Dinky" stores in the Omaha and Council Bluffs area; my uncle's store was the largest of the markets in this chain.

I was a bag boy; I worked at the check-out stand and packed the groceries into paper bags for the customers to take home. I was also supposed to be a carry-out boy; that is, to carry the filled grocery bags out for the customers and to help them load the bags into their cars. I worked a total of six hours on Saturday afternoon; I was paid 50 cents an hour. The six hours were interminable — they seemed more like 60 hours to me. Once in a while, but not often, I was rewarded with a small tip when I carried out the groceries for one of the customers, a nickel or a dime; I do not recall ever receiving a quarter.

Everyone who worked at the "Hinky Dinky" knew that I had the job and was the youngest of the bag boys because my Uncle Lou was the store manager, so they tolerated me. I was not as big and strong as the other carry-out boys, and I was certainly not as big and strong as the Iowa men who bought their groceries in our "Hinky Dinky." Not infrequently, there would be a Saturday sale of some large quantity item such as, for example, 100 lb bags of potatoes or 50 lb bags of flour. When I arrived at the store on such an occasion, I would find a large stack of 100 lb bags piled high, in a complex double criss-cross pattern for stability. When a customer wanted to purchase such a 100 lb bag of potatoes, he or she would inform the checker, after which one of the bag-boys or the customer himself would walk over to the mountain of 100 lb potato bags, roll one off onto his shoulder and carry it out to the customer's car. It looked easy, but what happened in my first attempt to follow their example was predictable. I collapsed under the weight of the 100 lb bag, fell to floor with the bag on top of me and was not able to move. The customer whom I was supposed to be helping laughed loudly, walked over and with one hand lifted the potatoes and with the other hand helped me up off the floor. He then carried the potatoes out to his car. That was indeed a long afternoon for me.

After much thought and soul searching while at work during one such Saturday afternoon, I screwed up my courage and decided that I would call Ruthie and ask her out for a date; I would ask her to go to the movies with me. When I called during my afternoon break and asked her and told her that I really wanted to go out with her and that we would see a Dean Martin and Jerry Lewis movie and that I heard this film was very funny and that we would have to take the streetcar into town and back, to my amazement, she said "Yes." We went to the Martin and Lewis movie that evening, laughed a lot, stopped for a coke afterward, and I took her home — and that was how our love affair started.

I have always been ambitious; I was certain that I could find a part-time job that was better than being a bag boy at the "Hinky Dinky" store. The following summer, I was successful in my search; I was hired as a stock boy in a woman's shoe store, Baker's Shoes, in downtown Omaha. What I really wanted was a job as a salesman to sell the shoes, but the manager, Mr. Isaacson, told me I was too young, that I had to be 16, and that he currently had all the help he needed on the selling floor. So I was confined to the depths of the basement where I sorted new stock when it was delivered and attempted to straighten out the mess of years of incompetence. Each and every shoe box had a code on the front; the code identified the kind of shoe (for example pump, sandal or bedroom slipper), the color, the size and the price. There was a special code for old styles; if a salesman was able to sell one of these old styles left over from earlier years, he would get a 25 cent bonus. When a particular style or size was depleted in the back room behind the selling floor upstairs, my responsibility was to bring up new stock from the basement and fill in the missing item. I remained in this stock-boy purgatory for approximately six months, working part time.

That summer, I rode my bicycle to Ruthie's house several evenings each week after work. Her parents, Harry and Bess Chudacoff, had a small grocery store which they proudly called the "Ames Avenue Food Shoppe." I guess we were all simply peddlers by ancestry, but the "pe" at the end of Shoppe was a statement of new and better expectations. Harry and Bess worked together all day in the Ames Avenue Food Shoppe, locked the doors at about 6:30 pm and then typically relaxed during the evening by going out for a leisurely dinner at one of their three or four favorite restaurants. I quickly became one of the family and often, especially during the summer months, joined Ruthie and her parents when they went out in the evening for dinner.

The Ames Avenue Food Shoppe was open on Sunday. As a result, on Sunday Ruth had access to her father's car. At that time in the state of Nebraska, one could obtain a learner's permit at the age of 15 and one-half. Harry taught his daughter how to drive so that she was ready to become a licensed driver as soon as she was 16. Ruth subsequently taught me to drive, in her father's Pontiac, on those Sunday afternoons. She taught me well; I passed the test for my driver's license, both the actual driving test and the written test, on my first attempt.

Ruth and I got into trouble only once. After I had learned to drive, successfully passed the driving examination and received my driver's license, we were out for a ride on a Sunday afternoon. I was driving. We were in the countryside near Father Flanagan's Boys Town, made famous by the movie with Spencer Tracy as Father Flanagan and Mickey Rooney as the bad boy ("He's not heavy Father, he's my brother"). Boys Town is located a few miles west of Omaha. I had my right arm around Ruth, and her head was resting comfortably on my shoulder. I was feeling very happy until we saw the flashing lights behind us. We were stopped by a Nebraska State Highway Patrolman. He gave us a serious lecture about the importance of paying attention while driving and instructed me in no uncertain terms to always drive with two hands on the steering wheel. He then gave me my very first traffic citation. The infraction, as written on the ticket, was stated as "Necking while driving." How I wish that I had saved a copy of that citation. What a gift it would have been for our grandchildren!

My days in the stockroom at Baker's Shoe store came to an end shortly after my 16th birthday. Without prior warning, Mr. Isaacson informed me one Thursday evening that he was giving me the opportunity to move upstairs. I had been promoted, and I would start greeting customers and selling shoes two days later, on Saturday morning. This was a real job. I made one dollar per hour plus a commission of 6% on all sales. The most expensive shoes in the store were the high-heeled dress shoes that sold for $6.99. Bedroom slippers sold for $1.99, and flats, sandals and sporty shoes were priced somewhere in between. One could make what to me was serious money from the 6% commission.

On that first day as a salesman, I was amazed how fast time went by. I was on my own, and the harder I worked, the more shoes I sold and the more commission I earned. This was a completely different experience from the drag of working for 50 cents an hour as a bag boy with no possibility for an up-side. I sold women's shoes during my last two years of high school and throughout my four years at the University of Nebraska. It is correct to say that I worked my way through university by selling women's shoes.

I quickly learned that the secret was to move fast; the most successful men on the selling floor were able to handle a number of customers simultaneously. I learned to bring out several boxes of different shoes in the size and general style requested by the customer, stack the boxes in front of her and open only the top box. Then I would put the right shoe from the open box on her right foot, tell her that it looked great on her (!) and run off to greet another customer at the front door. When I returned to the first customer a few minutes later, she had typically opened all the boxes and tried on all the shoes. If I was lucky, the sale was done. If none of the shoes was satisfactory, I returned these first selections to the back room, put them back into stock and brought out another stack of boxes, and so on. My success rate was reasonably high. On a busy day, I could handle three customers at a time, all day. The busiest day of the year was the Saturday before Easter. One such Easter Saturday, my best day ever, I sold nearly $2,000 worth of shoes and took home $125!

On slow days, I took my time during the lunch and dinner hours, but often somehow found my way across the street to Herzbergs, where Ruthie had gotten a job selling purses and handbags for women. She was only 15, but her bearing and beauty caused the management to believe her when she told them she was 16. She was comfortable and at ease with her customers. She looked great behind that counter, very mature and sophisticated.

After work and sometimes on lunch break, I spent many hours being "creative" at the Rocket Pool Hall. My specialty was snooker. I reached a reasonable level of proficiency, but I was never able to beat Charles Chisholm. Chisholm was older than me, already studying at the University of Omaha, and he "hustled" me every week on payday. Every Friday I was confident that I could beat him; every Friday my confidence cost me about $5. But I never lost my nerve; I was confident that one day I would beat him. On a good day I can still drop a few balls on the pool table and make it look easy; the game is after all only physics! Shortly before my 70th birthday, I challenged my young colleague at UC Santa Barbara, Tom Soh, a brilliant engineer who will revolutionize biotechnology. I played remarkably well and won the game of Eight Ball only by making two amazing triple bank shots in a row. Tom was impressed, but after that initial game, he beat me badly and regularly; the next time we played, he advised me that I needed to practice a little.

Our high school lives were centered within a group of clubs, clubs for girls and clubs for boys. The hangout was in the Jewish Community Center building across the street from Omaha Central High School. The Jewish Youth Clubs for high school teenagers in Omaha were a classic example of a class-based society. Some of the clubs were very exclusive and "snooty"; only the students from the more wealthy families were allowed to join and become members. This class-based society was to me the source of unpleasant memories for many years.

I joined the Jewish Youth club called AZA "1" at the start of my first year in high school. AZA stands for "Aleph Zadik Aleph," the first and last letters of the Hebrew alphabet. There were AZA clubs in cities all over the Midwest. The organization had actually started in Omaha. Ours was the founding chapter, hence we were AZA "1." We had regular formal meetings on Sunday afternoons at the Jewish Community Center. One had to learn Robert's Rules of Order and abide by these rules in the course of the meeting. "Point of Order" or "Point of Information" would be loudly shouted. The Sergeant-at Arms often had to attempt to restore order.

At one of the meetings, shortly after I had become a member of the organization, there was a discussion about the fact that AZA "1" did not have a proper fight song to support our sports teams, to make us proud and to generate *esprit de corps*. At the end of the meeting, I was determined that this was my opportunity — I would write the song for AZA "1." I took the melody from "You're a Grand Old Flag," wrote rousing new lyrics, introduced my song at the next meeting and sang it for the group. I handed out the words on a mimeographed sheet and convinced them to sing it along with me. It was a "hit" and immediately adopted by unanimous vote as the official song of AZA "1."

This initial success set me apart. I had taken the initiative and demonstrated that I could actually do something. I learned that action is the first step in leadership: Shortly thereafter, I was appointed Sergeant-at-Arms. In my sophomore year, I was elected as a senior officer and in my third and final year in high school I was elected President of AZA "1."

When starting out in any new endeavor, do something! Get the first accomplishment on record. I remembered this lesson on the fast-track to success when I arrived as a young Assistant Professor at the University of Pennsylvania. I quickly got my lab equipped and operational, and I carefully chose my first research problems. As with the AZA "1" song, my initial success set me apart; I was rapidly promoted through the academic ranks to Full Professor. At that time, I was the youngest Full Professor of Physics in any of the Ivy League universities.

I had a successful high school experience both socially and academically. I had a group of friends that was sufficient to fill my life. I was elected President of my Youth Club, and I had outstanding grades.

The wealth-based class structure did, however, diminish the quality of my high school experience. I never returned to Omaha Central High after Ruth and I married and departed permanently from Omaha. We returned only many years later, in 2002, when I was inducted into the Central High Hall of Fame, and then again on the occasion of my 50th Class Reunion one year later in 2003. At the 50th reunion, one of the girls, now of course an elderly woman, who had been perhaps the worst snob while in high school, confided to me somewhat sadly about how her life had turned out. She reminisced about how badly she had behaved during those high school years.

By taking extra courses, and by taking only the minimum required courses in literature and history, I graduated from high school in three years instead of the planned four-year program in the 9–12 system that was the norm at Omaha Central High. I was restless and wanted to get on with my life. I felt that somehow, in terms of my academic progress, I was not yet involved in the "real thing."

I took geometry and algebra in high school, but not trigonometry, and calculus was not even offered. I did not take a course in chemistry. The most important course that I took was physics; the physics course was not important because I enjoyed the subject, but, on the contrary, because it was an unsatisfactory experience. Our physics course was taught by Mr. Roy Busch. My only serious memory of that course is that I found physics, as he taught it, to be so murky that I later concluded that something was wrong — it had to be possible to understand physics. That unsatisfactory feeling is perhaps the reason that I am today a scientist.

During the spring of our second year in high school I asked Ruth to be my date for the annual Sweetheart Dance. This was a formal dance and a very special event. In preparation for the dance, I asked my cousin Milt Soskin for advice. What could I do to make progress with Ruth, to win her favor? Milt was 10 years older than I. He had been in the Navy during WWII, and he was one of the "lucky" ones who were called to serve again in the Korean conflict. He was very experienced, and he had many girlfriends. He responded that I had to "have a line." When I asked what he meant by that, he explained that I had to prepare words to say to her that would make her believe that she was very special. I thought about this advice at great length, but I had no idea what I could say to her as "my line."

The evening of the dance, when I came to her house in my dark suit, white shirt and carefully chosen tie, Ruth was wearing a strapless formal gown; she looked spectacular! I knew that she would be wearing a strapless gown — although I was not prepared for the effect that it would have on me — so I brought her a corsage of pink orchids that was designed to be worn on her wrist. Later in the evening as we were dancing, in the romantic atmosphere with the lights dimmed, I thought again about the need to "have a line." At some point, I simply told her that I loved her. In the years that followed, I have learned over and over again the power of those simple words. She believed me, and perhaps even more important, I believed me.

We were lovers at age 16, I gave her my fraternity pin when I was 19, we were engaged to be married when I was 20, and we were married on August 11, 1957 when I was 21 years old. Since Ruth was only 20, her father had to come with us to the courthouse to formally give his permission and thereby allow us to obtain the marriage license. He offered to pay the $2.50 fee, but I refused the offer; the fee — and his daughter — were my responsibility.

A few weeks prior to the wedding, my mother confessed to me that a year earlier she had added Ruth to her Blue Cross Health Insurance policy. She wanted to be certain that we were covered; the "pill" was not yet available.

I applied for admission only to the University of Nebraska. I had no concept of going elsewhere, for example to an Ivy League university in the eastern part of the United States, nor did I have the money to do so. Moreover, I was not about to move a long distance away. In fact, during my freshman year in Lincoln, I came home many weekends to spend time with Ruth. I often traveled the 60 mile journey from Lincoln to

Omaha as a hitch-hiker after working late in the day on Saturday. I rarely had to wait more than a few minutes before catching a ride to Omaha.

We went out on Saturday night, and as a result, I slept late on Sunday morning. I had Sunday lunch at home with my family with Ruth typically joining us, and took the train back to Lincoln late on Sunday evening.

Ruth joined me at the University of Nebraska the following year. Although we lived in separate "Greek" houses (she joined the Sigma Delta Tau sorority, and I had become a member of Sigma Alpha Mu fraternity), this period during which we were both away from home at the University of Nebraska was the beginning of our long life together.

That was more than 60 years ago. As it turned out, Ruth lived up to her biblical namesake; she and I have gone a long way together. Our love and Ruth's support are fundamentally responsible for all of our accomplishments including my success as a scientist, the award of the Nobel Prize, my becoming a successful entrepreneur, and our adventures in the theatrical world.

I have enjoyed the life of a scientist while sharing both the exciting days and the disappointments with her. She has filled my life with love and surrounded me with beauty. Her good taste in art resulted in the acquisition of many of our most interesting pieces, including our very first acquisition, from Hyde Park in London on a sunny afternoon in 1964, for the princely sum of $28. That painting, acquired 50 years ago, still hangs prominently in our living room. She has a natural creative talent; her original oil paintings decorate the walls of our house.

She has gallantly put up with my eccentricities and my workaholic tendencies. We have shared my life as a scientist in many ways. She often accompanied me to international scientific conferences where we built long-lasting friendships with colleagues and their wives. We have a rich "common memory" of these friendships and the many events that occurred at the series of conferences that span the 24-year period from the initial discoveries of conducting polymers to the award of the

Nobel Prize. These friendships enriched my science and became an important part of our personal lives.

Finally, and perhaps most important, we have succeeded in passing on my passion for science; our two sons, Peter and David, are creative and successful scientists and wonderful fathers to our four grandchildren.

7 Passion for a Career as a Physicist

I was the first in my family to obtain an advanced degree and one of the first to go to university. As my career as a scientist began to emerge, I thought that perhaps there had been a great Rabbi in my ancestry, someone from whom I could trace the genes, but that seems not to be true. My Grandfather Heeger was described to me as a "some-what crude" and uneducated man. The Heegers came from a town called Surhaz, in what is now Belarus, near the Russian border. The family emigrated from Russia by way of Odessa and immigrated into the United States through Ellis Island. Iowa was far from the centers of Jewish immigrant population in the eastern cities of the United States. Evidently, they were risk-takers; fortunately, I inherited those important risk-taking genes.

They arrived in the United States of America in 1904, when my father was four years old. My Grandfather Heeger had a sister who had married into the Rykoff family. The Rykoffs preceded the Heegers to Sioux City and, after arriving, in the tradition of immigrant families, helped to bring the Heegers to join them. Not long after Grandather Heeger brought his family to join his sister in Sioux City, the Rykoffs moved on to California and settled in Los Angeles. The Rykoffs thrived in Los Angeles. Again, they started out as peddlers, but grew their business into a large restaurant-supply company that eventually became a publicly owned corporation listed on the stock exchange. The familiar green Rykoff trucks displayed the motto "Enjoy Life, Eat Out more often," and were ubiquitous in Los Angeles and indeed all over Southern California for many years until the company merged into another and the Rykoff name disappeared from public view.

I know even less about the history of my mother's family and how they managed to get to the Midwest, to Omaha. I know only that my Grandfather Minkin worked for the Union Pacific Railroad. Perhaps the opportunity to find a job with the Union Pacific Railroad was the attraction that brought him to Omaha. My grandparents on the Heeger side died before I was born, as did my Grandfather Minkin. As a child, I knew and visited my Grandmother Minkin at her home in Omaha, but she was gone approximately a year after my father died.

There was no history of academic achievement in my family. I credit my motivation to go to university, to go on for a PhD and to eventually pursue the life of a scientist and a university professor to my mother. The inscription on her grave stone in the cemetery in Sioux City is simple and true: "Her legacy is the education of her children." And she did well; both of her sons thrived in the very special atmosphere of university life. My younger brother Gerald (Jerry) followed me into the academic world. He spent his undergraduate years at the University of California at Berkeley, completed his PhD studies in Political Science at the University of Chicago with an emphasis on South Asian studies and began his career as an Assistant Professor in Political Science at the University of Virginia. He subsequently moved into university administration and became Dean of Continuing Education, first at the New School in New York and then at New York University. He was subsequently recruited by and appointed as President of the University of Maryland University College. He is a pioneer, an innovator and a leader in the field of distance learning and online education.

I went to the University of Nebraska with the intent of studying engineering, in particular electrical engineering. In reality, I had no idea what was involved in such studies nor of what a career in electrical engineering would be like. I quickly found out, however, that engineering was not for me. With the exception of calculus, I was not enthusiastic about any of my first-year engineering courses.

Calculus was different; calculus is a revolution to the mind. Before calculus, there are a huge number of mathematical problems that one cannot even comprehend, questions that one cannot even ask. After learning calculus, one can actually solve such problems. This, I think, is where the passion began.

With that background (and the experience of a year of introductory physics at Omaha Central High), I registered for my first physics course. When Professor Saul Epstein walked into the room on the first day of class, I was truly surprised. He looked almost like a student. He was young, he was dressed informally with an open shirt, no tie and no jacket. He was the first physicist that I had ever seen and the first physicist in my life.

The class was large, with many pre-medical students. The pre-med students worked hard, they were intensely competitive and grade conscious, and some of them were very smart. As the year went by, I began to see through at least the elementary aspects of physics and to enjoy the development of the subject.

During the first semester, I did well on the mid-term examination and wanted to prove to myself and to Professor Epstein that I was a promising student. The final exam was difficult, but I finished it on time, and I thought I had done reasonably well. The custom was to leave a self-addressed postcard when you turned in your final exam booklet. After the exams were graded and the final course grades determined, the professor or his teaching assistants would fill in the final exam grade and the final course grade and mail the post card.

So I waited for the postcard. My "buddies" in the fraternity house knew I was concerned over the outcome, and they knew I was naïve. One of them simply took a postcard and without even attempting to match my handwriting sent the card to me with the information that I had failed the final exam and that I had received a barely passing grade in the course. I was devastated. Even though I could see that this was not the card I had deposited on the table with my final exam, I was truly devastated. I fell for the prank; I was convinced that my card had been lost or inadvertently torn, and that the teaching assistants had made out a new one. I was upset for 24 hours until the mail arrived the following day. There was my real card, the one I had addressed in my own hand! I had received the highest grade in the final exam and achieved the highest grade in the class. That night I truly celebrated — another perfect day!

There were many celebrations and many parties at the Sigma Alpha Mu House on the University of Nebraska campus in Lincoln, Nebraska. The fraternity house had a large living room with a smooth tiled floor and with an ample number of sofas and comfortable chairs, in various shades of brown, distributed about. On weekday evenings, the living room was typically quiet with "brothers" lounging, studying or often simply sleeping. On weekends, however, that large living room was the center of, and somehow the origin of, many memorable parties. Some were formally planned, but more often they were impromptu affairs that came together late in the evening after people returned from the movies or other affairs on campus. The sofas and chairs were easily and quickly pushed aside to the walls leaving a large open area for partying, for dancing and for listening to the music.

These impromptu parties would perhaps be better described as "jam sessions." A grand piano stood in one corner of the room, and the jazz musicians assembled around the piano.

The musicians who came together at the Sigma Alpha Mu House to play on those weekend nights were truly talented. Two of the best were Frank Tirro and Arnold ("Tuffy") Epstein. Both Frank and Tuffy played the clarinet and various kinds of saxophone. I had known Frank since grade school; he was a fat little kid who was born with a clarinet in his mouth. He was always the star at school concerts. Frank and Tuffy each made a life out of his music. Frank grew up (no longer was he a fat little kid!) and went on to graduate studies in music. He subsequently composed, became a specialist in the history of jazz (he wrote a definitive history of jazz) and the music of the Renaissance, and ended his long and productive academic career as Dean of the School of Music at Yale University. Tuffy took a different path. The "Tuffy Epstein Band" was a long standing feature on the Omaha scene. When I visited Omaha years later, I saw that Tuffy and his band were appearing at the lounge in the best steakhouse in Omaha.

When I returned to Omaha for my 50th Class Reunion in 2003, the Tuffy Epstein Band played at the Reunion Dance for all of us "old folks." Even better, Frank Tirro was there, and the two of them played together again for the first time after nearly 50 years. They were terrific; there was not a dry eye in the audience.

There was never any written music at these jam sessions; the guys just played and played long into the night. Plenty of beer was available, and nearly everyone in the house had a stash of some kind of hard liquor. As far as I know, however, drugs were not yet part of the scene. No one ever smoked marijuana at our fraternity house. People came and went and eventually drifted off after 1 am because the girls were required to be back in their dormitories or sorority houses before the scheduled lock-out time, but the music went on.

The jazz in Lincoln was terrific in the 1950s. Even Louis Armstrong came to the University of Nebraska — two times while we were there. The Louis Armstrong Band played at the spring dance two years in a row, during my junior year and again during my senior year (1956 and 1957). Armstrong was of course already a legend, and to see him in person and dance to his music were indeed special experiences. Both times, Ruth and I managed to find our way up to the front, right near the stage. That remarkable face, that signature grin and that amazing voice — the man had style and he had class.

I completed my undergraduate studies with a dual major in physics and mathematics. While doing so, I continued to aspire to a leadership role; I served as Treasurer and then President of the Fraternity. I had not yet made the personal commitment to dedicate myself to a career in science.

In mathematics, I went through the typical range of courses, calculus, linear algebra, differential equations and mathematical analysis, such that I had a reasonably strong background in preparation for graduate school. In physics, however, I felt that I was less well prepared. I had taken the standard courses in classical mechanics and electricity and magnetism. The physics courses seemed dry and old to me; part of history rather than a living and exciting subject. As a result, I was on a track to go on to graduate school with the intent of working toward a PhD in mathematics.

My plan to become a mathematician was abandoned during my senior year as a result of a course in modern physics taught by Professor Theodore Jorgensen. Jorgensen introduced me to quantum physics and 20th century science. Even today, much of what I really know about the fundamentals of modern physics can be traced back to his course. We started with black body radiation and Max Planck's seminal concept of quantization. He introduced us to Einstein's three important contributions from the famous Great Year of Physics (1905), the Special Theory of Relativity, the Photoelectric Effect, and the Theory of Brownian Motion. We studied the Bohr atom and the beginnings of quantum mechanics, and we studied atomic physics more generally. He concluded the course with an introduction to nuclear physics. I loved the material; finally I had the sense of a living science that was progressing toward answering new, important and interesting questions. Jorgensen's course convinced me that I should seriously consider a career in physics.

I went one day to Professor Jorgensen's office to ask him for advice. I had little confidence that I could actually succeed in a career as a physicist. The question that I wanted him to answer was difficult for me to articulate, but I needed to understand better what was required for success as a physicist. He thought for a moment, but did not respond as I had expected. He simply asked me how well I had done in my English courses. I was surprised by the question, and I expressed my surprise to him. He commented that of course success in physics required that one be dedicated, motivated, creative and smart, but that the ability to communicate was of critical importance. One had to be able to write well and clearly, one had to be able to express himself and his passion for science clearly. Professor Jorgensen emphasized that even important scientific progress would have little impact unless the scientist was able to communicate his discoveries to his colleagues in the scientific community.

I was recently reminded of Jorgensen's comments and advice when I read a historical account of the great discovery of penicillin by Alexander Fleming. In 1928, Fleming discovered, quite by accident, that this peculiar mold had the effect of killing microbes. He did good solid work; there was no mistake or uncertainty in his conclusions. Although he reported his findings to the Royal Society in London, his style and delivery were so poor that no one paid any serious attention. He more or less spoke into the inside pocket of his jacket with little or no emphasis. Fleming's written papers were evidently equally boring. He might have had passion for science, but he was not able to express it.

Ten years later, through the efforts of Howard Florey, Ernst Chain and colleagues, Fleming's work was rediscovered. Moreover, because of the style, the persistence and the charisma of Florey, penicillin emerged as the "Miracle Drug." The Age of Antibiotics actually began during the Second World War in the early 1940s. Fleming, Florey and Chain were awarded the Nobel Prize in Physiology and Medicine in 1945 for their discoveries that resulted in the widespread use of penicillin. Before penicillin, people died from infection from even the simplest wounds, such as from a scratch by a thorn on a rose bush. One wonders how many lives might have been saved if Fleming had been a better communicator in 1928, such that the development of penicillin would have occurred on a shorter time scale.

Ted Jorgensen and his course in modern physics instilled in me something else that has lasted through my entire life: He instilled in me a passion for physics. Passion is important for any endeavor; passion is necessary for overcoming the many obstacles that stand in the way of success. Great musicians are passionate; one can see the passion in their work and one can feel the passion when listening to their music. All creative people are passionate about their work; success in any field that involves creativity requires passion. The "ups and downs" cannot be tolerated without the underlying passion that drives the creative person onward.

One might be surprised to learn that scientists are truly passionate about their research. A picture has emerged in our culture that science is "dry" and even boring. To a scientist, however, obtaining new data is exciting; the idea for the next experiment or the insight into understanding a result obtained from an experiment is fulfilling at a very deep level. Great scientists are passionate about their research. That passion drives them forward day after day in their endeavor to uncover the secrets of Nature.

Not infrequently, I awake in the middle of the night, open my computer and spend the next few hours working on a manuscript, struggling to get words precisely right and to make the meaning absolutely clear. When I was younger, all-night writing sessions were not uncommon. Why? Because it was, and still is, important to me to be the first to understand the physics underlying a new result, and to be the first person to get the science right. That need to be both first and right is an expression of the passion of a scientist.

I was honored by the University of Nebraska in the late 1990s with the award of an honorary degree, Doctor of Science Honoris Causa. On that occasion, I had the pleasure of giving a Colloquium Lecture to the physics department. Ted Jorgensen came to the lecture; he was 92 and long retired. He looked the same to me, and he proudly told me that he was working on a revised edition of his successful book on *The Physics of Golf*. I told him how much his course had meant to me, and I thanked him for his good advice. Professor Jorgensen celebrated his 100th birthday in 2005. I am certain that he influenced many other young minds just as he influenced me. During my many years as a professor, I have often repeated Jorgensen's advice on the importance of developing one's communication skills to aspiring students in any field of science. I gave similar advice to each of my sons. I have heard each of them lecture and I have read some of their publications. They learned well; their lectures are clear and they know how to engage an audience. Their written publications do not suffer from the dullness of those written by Alexander Fleming. They, too, have passion for their science.

I have advised my grandchildren and many young people about the need to find a career path about which they are passionate. Without passion, success is impossible.

8 **From Novice to Professor**

Ruth and I were married in the summer of 1957, and we departed from Omaha to start our life together, and to start my graduate studies. We departed without fear although we had nothing, no money — with only a naïve optimism and my undefined passion to go forward toward becoming a scientist. We left Omaha with a sense of excitement for our future together.

Throughout the 64 years that we have been together, our love has carried us very far. Our love is responsible for what we have become. Our love and her support are fundamentally responsible for all of our accomplishments including my success as a scientist, the award of the Nobel Prize and my becoming an entrepreneur. Clearly, and again for reasons that I do not deeply understand, she shared my passion.

I find it amazing and remarkable that Ruth and I have come so far and remained together. We were very young and unsophisticated when the odyssey began. We had no world experience, having grown up in Omaha without the benefit of extensive travel. Yet we somehow managed to grow up together, and to change together — rather than drift apart as so often happens — and to stay together.

I had applied for admission to the graduate programs at several universities, but decided to accept the offer to go to Cornell University in Ithaca, New York. Cornell had a great tradition in physics and a strong and large graduate program. During the decade following World War II, Cornell put out more PhDs in physics than any other institution in the United States. I wanted to test myself against the highest level of competition. At Nebraska, I had made a number of friends who were clearly very smart, and who seemed to pick up new concepts in mathematics and physics with remarkable ease. But I was not yet calibrated; I did not know if we were really competitive on a world-class scale. Were we just a group of "hicks" from Nebraska? Thus, I aspired to go into a graduate program where I would be challenged by the best. I had applied to Cornell, Stanford, the University of California at Berkeley and Caltech. Stanford and Caltech turned me down. The decision to go to Cornell rather than Berkeley was determined primarily by the desire to experience one of the great Ivy League institutions.

The day after our wedding, we pointed our car toward the east and departed from Omaha without looking back. Our treasury comprised the $628 we had received as wedding gifts. Our honeymoon trip was the journey to Ithaca. We drove to Ithaca via Chicago, Toronto and Niagara Falls. We actually went to Niagara Falls for our honeymoon! The entire trip was an adventure. In Chicago, we went to an expensive night club to see a performance by Sammy Davis Jr.; at that time his father was still dancing and performing with him. In Toronto, we ate dinner at a wonderful French restaurant. At Niagara Falls, we did what newlyweds are supposed to do — and we also took a boat up the river to see the Falls from below.

We arrived in Ithaca on a sunny Sunday afternoon. We found a furnished apartment in an old house up on the hill above Lake Cayuga across the gorge from the main Cornell campus. I could easily walk from the apartment to Rockefeller Hall, the physics building. Ruth found a job in the College of Agriculture. She was first offered a job in the entomology department, but turned it down because she does not like insects. Her job in the veterinary school involved chemical analysis of cattle feces; even that was better than insects. Although her job did not pay very well, it was sufficient when added to my annual stipend of $1,300 to allow us to get started.

We spent the first few weeks exploring Ithaca, Lake Cayuga and the Finger Lakes region. In this upstate New York area, the waterfalls are plentiful and truly beautiful. We walked along the famous deep gorges that cut through the Cornell campus. It was an idyllic period.

Two events occurred that brought us quickly back to reality. First, in late September, classes began. Second, in late October, winter began. The leaves quickly fell to the ground to form a thick brown carpet, the weather turned cold, snow began to fall, and the sun disappeared and was not seen again until the following April. The change of weather was perhaps fortunate, for I had a serious amount of work to do.

That first semester at Cornell was a turning point for me. I learned that I could compete. There is a theorem, proven by experience, that one can always find someone who is smarter than you. But in fact, I did well. I learned a great deal in class, but I learned even more from my interactions with fellow graduate students. To my surprise, most of them were as naïve and unsophisticated as I was. We solved difficult physics and math problems together. We managed to get through Cornell's infamous experimental physics course. We discussed endlessly the various options: Should one work toward a degree in nuclear physics — or solid state physics?

In 1957, everyone knew about nuclear physics; the development of nuclear weapons was still much in the news and in the public mind, and Cornell had an especially strong program with a state-of-the-art particle accelerator and associated facilities. Hans Bethe was at Cornell, and the legendary stories of Richard Feynman and Freeman Dyson at Cornell were often repeated. Although I had taken a course in elementary solid state physics during my senior year at Nebraska, I was not seriously aware of opportunities in solid state physics. However, I quickly came to realize that the invention of the transistor was the beginning of a revolution in electronics, and I learned a little about the wonders of the many-body physics of superconductivity and of magnetism. Equally important, perhaps even more important, the Russians had launched Sputnik; the so-called "Space Race" had been initiated. It seemed that there would be many job opportunities in solid state research for young physicists, opportunities both in industrial laboratories and in universities. Although I still knew very little about the actual subject matter, I decided that a career in solid state physics was the correct choice.

At the end of our first semester in Ithaca, we accepted an invitation from one of my graduate student colleagues to visit him, his family, and his fiancée in Philadelphia during the holiday period. Ruth and I drove from Ithaca and stayed with them for a few days. We enjoyed their hospitality as well as our first introduction to the culture of a large city on the East Coast. Philadelphia did not make a great impression on us. Little did we know that we would return almost precisely four years later for a 20-year period during which I would initiate the research at the University of Pennsylvania that eventually resulted in the award of the Nobel Prize.

After a few days in Philadelphia, we moved on to New York and started our love affair with that wonderful city. We stayed in a very old hotel owned by some relatives of a friend of Ruth's parents. It was little better than a "flop house," but it was cheap, supposedly clean, and the location was terrific; on 48th Street near Sixth Avenue. We looked up at the tall buildings, we went to the Automat, we went to museums, we went to Greenwich Village, and we explored for the first time the foods of many different cultures.

We went to the theater, to the Broadway stage, for the very first time, to see *Two for the Seesaw* with Anne Bancroft and Henry Fonda. That performance was our memorable introduction to the great tradition of New York theater.

In my wildest imagination in 1957, I would not have anticipated that five decades later, we would invest in and participate in the production of Broadway plays. Our first production was a revival of Neil Simon's *Barefoot in the Park*. The opening night performance was terrific. The cast was strong, their timing was perfect, and we laughed as Neil Simon intended that we should. It was a wonderful night at the theater. Then, just as in the many old Hollywood films about the Broadway theater, we waited for the reviews at the party following the opening night performance. To our dismay and disappointment, the review in the *New York Times* was one of the worst I have ever read; not just bad, but vitriolic! Investing in the theater is a different kind of risk than any in my experience. In contrast to being determined by measurable accomplishment, the fate of a Broadway production often depends on the qualitative opinion of a single reviewer. Fortunately, our involvement in New York theater did not end with *Barefoot in the Park*. We have gone on to participate in several additional Broadway shows:

- *In the Heights* — a musical that was awarded seven Tony's and enjoyed a two-year run on Broadway followed by a lengthy road tour.
- *West Side Story* — a revival of the famous musical by Leonard Bernstein. This, too, enjoyed a two-year run followed by a lengthy road tour.
- *Cinderella* — a musical by Rodgers and Hammerstein (both long dead) that had never been produced on Broadway. *Cinderella* opened on March 2, 2013 and closed January 3, 2015.

Thus, our love of the theater has resulted in a minor professional involvement that continues to this day.

We had a wonderful time in New York in December of 1957, a memorable time, and we spent all our money. We actually planned our exit very carefully, and we almost made it. We had retained sufficient funds to purchase gasoline for the trip home from New York City to Ithaca and to be able to eat a little along the way. We forgot about the toll road. When we realized that we would have to pay toll at our exit, we examined our joint resources and found that we had little to offer. I had two precious silver dollars that I had carried in my pocket since our wedding day, nada mas. I suggested that I could offer the man in the toll booth my Timex watch. Under such circumstances, the best policy is always simple honesty. The man in the toll booth laughed at our predicament and at our naiveté. He told me that he did not want my silver dollars, nor did he want my watch. He gave me an envelope with instructions to mail the few dollars that we owed to the State of New York.

During our period in Ithaca, we mostly ate chicken pot pies. Amazingly, we could buy frozen chicken pot pies at the local supermarket for 10 cents each. I thought this was a great bargain, and I convinced myself that I really liked chicken pot pies. Ruth never quite agreed. She is a butcher's daughter, and she liked a good steak once in a while; truth be told, so did I! Nevertheless, after returning from the trip to New York, we ate a lot of chicken pot pies.

We decided to leave Cornell after one year because we did not have the money to continue my graduate studies there. Ruth was pregnant with our first son. Without her salary and with the expectation and responsibility of a new baby, trying to continue toward the PhD in physics at Cornell just seemed much too difficult.

The year at Cornell was a great experience. We grew up a little, we found out how to survive on relatively little, the world began to open up for us, and we began to acquire some good tastes in culture, especially theater. For me as a young scientist, I gained the confidence that I needed, and I began to acquire some of the knowledge and sophistication that were required to go forward.

I had heard about interesting opportunities in California. The Lockheed Space and Missiles Division and Hughes Aircraft offered programs which allowed one to work part time while simultaneously continuing part time in graduate school. The company would even pay the tuition costs. This seemed like a possible solution to my problem, so I applied and was accepted into the Lockheed program. I would receive the truly remarkable salary of $6,600 per year while in this special program!

I was informed, however, that acceptance into graduate school was my responsibility. There were two suggested options: physics at Stanford University in Palo Alto or physics at the University of California at Berkeley. I applied to both, but Stanford was clearly preferable simply because Stanford was right there in Palo Alto whereas Berkeley was

50 miles away. I was accepted by Berkeley; Stanford, once again, turned me down.

The following year in Palo Alto was remarkable in many ways. Our first son, Peter, named of course after my father, was born on October 19, 1958 in the Redwood City Hospital. I presume that all first-time fathers with male offspring have the same memory of the first sight of their newborn sons. Peter appeared in Ruth's arms as they wheeled her out of the delivery room. She smiled and glowed and looked a little tired; he was beautiful, and he looked just like me!

Peter was a great little guy, and, of course, he did everything better and faster than any of the children of our friends. He stood up early, he walked early, he began to talk early; he was our little prodigy.

We lived remarkably well on my princely salary of $550 per month. We had a spacious apartment, made new friends and enjoyed Palo Alto, San Francisco and the San Francisco Bay area. We ate out relatively often, hired baby-sitters and took in films, and took advantage of some of the many cultural activities around Stanford and in San Francisco. In spite of this new and enhanced lifestyle, we managed to save a lot of money (at least in our terms). These savings turned out to be critically important; they were sufficient in fact to get us through the final three years of graduate school.

During the year at Lockheed, I worked on a solar cell project — silicon solar cells for use in satellites. Silicon solar cells had been discovered at the Bell Telephone Laboratories in 1954, only four years earlier. Although the initial announcements by the Bell Labs executives had promised unlimited energy in the future, in fact the technology took off very slowly. Silicon solar cells were very expensive and they were not very efficient. The onset of the "space race," however, provided the ideal first significant application for solar cells, since the batteries in Sputnik and the early satellites discharged in a few days or weeks. Solar cell panels provided the solution, a long lasting renewable source of energy. There is plenty of sunlight in space, and the cost of the (at that time) expensive solar cell panels was insignificant compared to the cost of building, launching and maintaining a satellite.

Lockheed Space and Missiles Division had a major defense contract to construct and evaluate solar cell panels for use in satellites. Although I did not realize in 1958 that I was at the forefront of the revolutionary use of solar cells in space, my job at Lockheed was interesting and there were a number of top-class people with whom I worked and from whom I learned a great deal. I filed my first patent with the U.S. Patent Office during my year in the Lockheed program.

Research on "plastic" solar cells fabricated with semiconducting polymers rather than silicon has been a major topic of interest for me for the past 20 years. Of course, when I was working on silicon solar cells for space applications at Lockheed in 1958, I had no idea that the plastic solar cell would play an important role in my future.

There were two reasons why I decided to leave the Lockheed program and to return to my graduate studies full time. First, I realized that if I did not do so, I might well be "trapped" by the responsibilities of the job plus the responsibilities of family life into an unsatisfactory career, a choice that I would certainly regret. Second, I realized that if I continued on the part-time path toward a PhD at Berkeley, I would not live very long.

The life threat is easy to understand. On Monday, Wednesday and Friday during the year at Lockheed, I set the alarm, awakened early, and drove the car on the Bayshore Freeway and across the Bay Bridge from Palo Alto to Berkeley to attend classes. After sitting in class all morning, I had lunch and then got back into the car and drove on the freeway again to return to work in Palo Alto. Naturally, after such a sedentary morning followed by lunch, followed by driving in the warm afternoon sunshine, I fell asleep at the wheel during almost every trip.

The sense of being trapped deserves more comment. The program at Lockheed was designed with the intent of finding bright young people and bringing their talent into the corporation. For those who entered this program with every intention of completing the work for their PhD, children came, expenses increased, houses were purchased and, in the final analysis, the realities and responsibilities of life determined their plans. Most of those who entered this program stopped their studies with the completion of the Master's degree. I take credit for seeing through this almost inevitable process and for realizing the implications of the combined pressures and the implied conflicts of work, family and graduate studies.

Thus, it was not a terribly difficult decision; Ruth and I agreed that completing my PhD studies was our highest priority. I resigned from the Lockheed program, we moved to Berkeley, and I started research on a full-time basis.

When I started at Berkeley, my goal was to do a PhD thesis in theoretical physics under Charles Kittel. Professor Kittel was the leading theoretician in solid state physics at Berkeley. He was famous for his book, *Introduction to Solid State Physics*, a book which educated an entire generation of solid state physicists. Thus, when the decision was made to go for my degree on a full-time basis, I went first to Professor Kittel and asked if I could work with him. Kittel had just returned from a trip to Moscow where he met the great physicist Professor Lev Landau, and he told me that Landau required that a prospective student must pass a rigorous examination before he (Landau) would agree to take that student into his research group. In that spirit, Kittel indicated that I should take the PhD qualifier exam and come back to him after I had done so. I passed the exam, and when I came back to discuss my future with him, Kittel told me that he would take me on as a research student. He said, however, that although I could do a thesis under his direction in solid state theory, he did not think I would be a first-rate theoretician. He recommended instead that I consider working under the direction of one of his colleagues who does experimental work in close interaction with theory. I was fortunate to have good advice at several critical stages in my life, and I was smart enough to listen to that advice. I followed Kittel's recommendation. I joined the research group of Professor Alan Portis.

I remember with clarity my first day in the laboratory. I was doing "original research"; at last I was involved with real physics. This was fun, it was exciting, and it was the "real thing." It was the realization of the passion that I had been carrying in my head for years.

After only a few days of carrying out magnetic measurements on an insulating antiferromagnet with the chemical formula $KMnF_3$, I had the audacity to write a "Theory of Antiferroelectric Antiferromagnets" and to present it to Professor Portis with great pride. He was patient with me then, and again a few days later when I apologized and told him my "theory" was nonsense. Through my interactions with Portis — I recall spending many hours talking with him in his office — I learned how to carry out research in physics. More important, I began to learn about good taste in the choice of problems, a subject that has continued to fascinate me throughout my life as a scientist.

This was a wonderful time to be a young scientist, to be starting a career in science. The post-war love affair with science, the dream that science would directly benefit large companies such as AT&T, Westinghouse, RCA, GE, IBM (and many more), and the reaction in the United States to the launching of Sputnik all resulted in many opportunities for young scientists with interest and expertise in solid state physics. This unbridled enthusiasm resulted in some bizarre events. For example, approximately two weeks after I had begun my thesis research in Alan Portis' laboratory, I was sitting at the lab bench puzzling over some small problem with the instrumentation. I heard a knock on the door and a well-dressed man wearing a suit and tie came into the lab and introduced himself to me. He said he was from the General Electric Central Research Laboratory and that they were interested in recruiting promising young scientists. I was astounded. It was like a scene from a Woody Allen film: Who me? I had not yet accomplished anything! This was indeed a wonderful time to be starting a career in science.

The years at Berkeley were not without stress. There were hurdles, such as, for example, the Advancement to Candidacy Examination. Although this exam was in reality quite routine, I took the challenge much too seriously and, evidently, I transmitted my concerns about getting past this hurdle to Ruth. Whether or not the resulting stress was the origin I do not know, but two months before the scheduled examination, she suddenly, and without warning, had a serious incident with a bleeding ulcer.

This was a very bad scene; but we did not lose our nerve. It was nearly midnight, and she was losing blood at an alarming rate! We lived in graduate student housing well away from the Berkeley campus. Our small apartment in the student housing area was located at the corner of 8th Street and 9th Street in Albany, California. I still do not understand how it is possible that 8th Street and 9th Street intersect; perhaps it is a physics or mathematics "joke" based on the non-Euclidean geometry that enabled Einstein to solve the problem of General Relativity. Whatever the reason, it was a confusing address. Thus, after I called the Emergency number at midnight on that moonless night and requested that they send an ambulance, the result was utter confusion. Eventually they found us, but only after I went out and "hailed" the ambulance as if it were a taxi on the street in New York City.

After a period in the hospital and some time to rest and recuperate, Ruth recovered, I passed my Advancement to Candidacy Exam without difficulty, and our life went on. But I remember that night more as a nightmare than as a sequence of real events.

The day following the ulcer attack, I realized that I needed help; there was no one to take care of little Peter while Ruth was recovering. Life has a way or repeating itself. I called Ruth's sister, Phylis Nerenberg; Phylis and her family lived in the Los Angeles area. Of course, without the least hesitation, she offered to help and to take care of Peter. He and I flew to Los Angeles airport the following day, and I sadly left my little two-year-old son there in her care. Peter was not away from us for very long, perhaps only a couple of weeks. Nevertheless, I re-lived the experience I had witnessed years before with my little brother; when Peter was returned to us by his Aunt Phylis, he addressed her as "Mommy" rather than Ruth. Peter quickly remembered his real Mommy and life went on.

By this time I was well into my thesis research, and I was obtaining interesting new results from my experiments, minor discoveries, but discoveries nonetheless. These new results drew the attention of my fellow graduate students, especially those who were young theoretical physicists looking for new results and new research problems to understand in terms of basic theory.

I recall my pride and the pleasure that I experienced when Philip Pincus first walked into my lab and told me that he had heard about my results and that he wanted to learn more. Phil was an advanced graduate student working toward his thesis under the direction of Professor Kittel. He was smart and had learned that keeping up with the latest experimental results gave him an advantage over those who did not. Over the next few months, Phil and I developed a collaboration and a friendship that has lasted many years.

In 1959, Phil was huge; he weighed more than 300 lbs. But he was truly a "jolly fat man." He was irreverent and poked fun at almost everything (as he still does today). Phil and our little son Peter became friends and "buddies." Phil often called on the telephone in the evening or on weekends and asked to speak with Peter. They would go on and on; somehow he found subjects to discuss with a two year old. Although Peter spoke clearly — and often — his vocabulary was limited. Phil taught him many things, including enough swear words to embarrass a longshoreman. On one weekend trip to San Francisco, Phil taught Peter how to say the name of the city of our destination. Peter learned to count before he was three so he knew all the numbers. Under Phil's tutelage, San Francisco became Seven-Six-Oh. Peter referred to that great city as Seven-Six-Oh for a number of years, well beyond the time when he had learned the correct pronunciation.

Phil went to France for a post-doctoral appointment, and the returned to an Assistant Professor position at UCLA. Although we remained in contact and continued to collaborate and publish together during my early years at Penn, I was not aware that he had gone on a serious diet. When we met at the meeting of the American Physical

Society in Chicago in the mid-60s, he had reduced in size by half; he lost nearly 170 pounds! Friends who had known him for years (including me) only recognized him after hearing his voice.

I completed all the requirements for the PhD in July 1961, three years after I had come to Berkeley and only four years after I completed my undergraduate studies. The transformation from novice to physicist took place over a very short period of time. I had no difficulty arranging interviews for possible job opportunities. I organized a job interview tour during mid-May of 1961 that took me by jet — my first time on a Boeing 707 — on the following route:

— Physics department at the University of Chicago
— Physics department at the University of Pittsburgh
— Physics department at the University of Pennsylvania
— Bell Telephone Laboratories in Murray Hill, New Jersey
— Raytheon Central Research Laboratories in Waltham, MA

I came home exhilarated and enthusiastic with all that I had experienced. I told Ruth that I had decided that an academic career was the best choice for me. I described how beautiful it had been in the Philadelphia area during my visit, and I told her about the wonderful hospitality that I had enjoyed during my visit to the University of Pennsylvania. Eli Burstein and Herb Callen were senior professors at Penn; they were building a major program in solid state physics. Eli and Herb had gone all out to recruit me. Donald Langenberg was a young Assistant Professor at Penn. I had previously met Don at Berkeley during the final months of his thesis research when I first joined the Portis Group. He had impressed me as a mature and accomplished young scientist; he had completed his thesis research when I was just beginning. Don and his wife Patricia were our age, and after spending one evening with them, I knew we would be friends. In any case, as I told Ruth, if I accepted the position at Penn, we would only be there for a few years before moving on, so she should not worry about being far from her parents and her sister and family. Thus, we made the decision to go to Philadelphia where I would become Assistant Professor of Physics at the University of Pennsylvania.

I found out later that I had visited the Philadelphia area on the two most beautiful days of the year. It was spring, the azaleas were in full bloom, the green was *very* green, and the colors of the flowering trees were spectacular. Eli Burstein had taken me on a trip through the Main Line area where many of the Penn faculty lived. He pretended to have lost his way, but purposely exposed me to the back roads in that beautiful area. No matter; we went to Penn and stayed there for 20 years, in many ways the best years of our lives.

I stayed on at Berkeley as a post-doctoral research fellow for six months following the completion of my PhD. During this period, Alan Portis took a sabbatical leave in Japan, Thus, I had my first small experience with "directing" a research group; the younger students looked to me for advice.

In September 1961, my brother Gerald came to Berkeley to start his undergraduate studies. Jerry is seven years younger than I, so we had been apart during the critical high school years when he was growing up. We got re-acquainted. We often met for coffee in the morning when he had a break between classes. Since I had not taken part in the formal graduation ceremonies, on one such morning coffee break, I told him that I had received notification that I should come to the Office of the Dean of Graduate Studies to pick up my PhD diploma. It was an amusing scene. After all my trials and travails, they could not find the diploma certificate. Eventually the official document was located and handed over to me. Now it was official; now I was truly a physicist.

The last event that occurred while we were living in Berkeley was one of the most important. Our second son, David, was born on October 3, 1961. We were all very happy, Ruth was happy and proud of her second son as well as looking forward to the beginning of a new life as the wife of a young professor at Penn, I was happy for all the obvious reasons, and Peter was happy because he was a happy little three year old, and he had not yet begun to understand the disruption that having a new little brother would cause in his life.

Peter and I went out to dinner together that night in Berkeley; I smoked a big cigar as a proud father was supposed to do, and he enjoyed the grown-up role of going out alone with his dad.

Was I prepared with my new PhD to immediately take on the multifaceted responsibilities of a young professor in a great academic institution? I did not really know, but I was confident that I could do so, and I was determined to give it my best effort. I went directly from graduate student to Assistant Professor without the benefit of a post-doctoral period to mature and expand my horizons as a physicist. Again, it was a risk, but a risk that was to me certainly worth taking. So with confidence and a little apprehension — off we went to Philadelphia and the University of Pennsylvania.

9 The Pathway to Great Science

Becoming a Scientist at the University of Pennsylvania

We arrived in Philadelphia; we moved from the University of California at Berkeley to the University of Pennsylvania at the end of December 1961. I arrived by car on December 27, 1961 having made the trip alone by car; Ruth came by plane a few days later. We made an intermediate stop in Omaha for several days to see our families. The first leg of our journey went smoothly, although Peter, now three years old, was somewhat stressed while in Omaha. He asked me many times: "Where is my bed?" Without his own bed, he had no real home. I explained to him that his bed was in a big truck and was being carried by that truck to his new home in Bala Cynwyd, Pennsylvania. My reassurances that he would soon have his own bed again only partially satisfied him. He remained dubious and unsettled, and he had no idea what Bala Cynwyd was or where it was located.

The trip from Omaha to Philadelphia turned out to be more difficult. We were to stay with our new friends, Pat and Don Langenberg. They were as remarkable then as they are today. They welcomed us with warm hospitality and took us in as part of their family. They even looked for and found for us, prior to our arrival, a house to rent so that we could settle into our new life with a minimum of effort.

The day after I had arrived in the Philadelphia area after my long trip from Omaha alone by car, Ruth departed from Omaha by plane with the two boys on what was supposed to be a non-stop flight to Philadelphia. Bad weather, a combination of fog and snow, in Philadelphia caused their plane to be diverted to Pittsburgh. They were eventually brought by bus on a long and harrowing overnight ride to the Philadelphia airport, where I greeted them. The never-ending bus ride was particularly difficult for Ruth with both a three-month-old baby and a three-year-old in her care. After arrival and transfer to the Langenberg residence, she slept for 24 hours.

We moved into our new home a few days later. The furniture arrived, including Peter's bed, and we began to get settled. We owned relatively little furniture, but filled in what we needed during the coming weeks, including a big comfortable chair that I purchased from the Salvation Army Store on Market Street, a picnic table with two benches which we purchased from Sears and used for a dining room table and finally, some months later, some living room furniture that we found on sale in an upscale Danish modern furniture store in central Philadelphia.

Not more than a few days after we moved into our side of the duplex on Greywall Lane in Bala Cynwyd, I began to go into the university every day to start the process of setting up the instrumentation for my laboratory in the physics building at Penn, the David Rittenhouse Laboratory. Every time we move into a new home, Ruth complains and teases that as soon as I have mounted the paintings and the artwork on the walls, I declare myself settled and then return to the regular routine of my life as a scientist, leaving her to complete the move. Since we had neither paintings nor artwork when we moved into the duplex on Greywall lane, I went off to the laboratory almost immediately.

For the first two years of our life in the Philadelphia area, we had only one car. As a result, I usually went to the university by public transportation. I waited for the bus at the bus-stop located two blocks from our house on City Line Avenue, went by bus to the 69th Street Terminal of the Market Street Subway line, transferred onto the subway, got off at 34th and Market and then walked about three blocks to the Rittenhouse Laboratory. Usually the trip went smoothly, but sometimes, either because of bad weather or because of what appeared to be random delays, it seemingly took forever. In all honesty, however, my biggest problem during that commute was getting past the doughnut shop in the 69th Street Terminal without stopping for a doughnut and coffee. These were the best glazed doughnuts I have ever eaten, and they remain unmatched in my extensive and lifelong search for the perfect glazed doughnut.

My style is to focus and to focus tightly, to take risks and to reach the forefront of knowledge as rapidly as possible. During the three years of my thesis research as a graduate student at Berkeley, I had focused on a relatively narrow area within the more general field of magnetism, which is itself a small part of the broader spectrum of physical phenomena known collectively as solid state physics (solid state physics has been properly renamed in recent years as "condensed matter physics").

Thus, I thought long and hard about what I would choose for the initial research projects in my new laboratory at Penn. The criteria upon which the choices were made were straightforward. I would not continue the research that I had done as a graduate student; I must start something new. One research thrust had to be what I thought of as a "sure thing," that is, a set of experiments that would result with high probability in publishable results. A second research direction should be a "high risk" venture; an attempt to make progress in an area that if successful would have high impact. The final criterion was that the instrumentation had to be relatively simple. I was not strong or talented in the development of new experimental techniques nor did I have the time to devote the several years that would be required to do so.

At this point in my career, I simply had to get going. I needed new results to publish and to demonstrate to my colleagues that I had real potential. Just as with the AZA song in high school, I needed something to set me apart. The important thing was to accomplish something, to get going. The rest would follow.

With the start-up funds available to me, I purchased a large electromagnet, the components for a versatile electron spin resonance spectrometer, and the components for a relatively simple radio-frequency bridge that covered a broad spectral range from a few megahertz to nearly one gigahertz. With the latter I would attempt to detect the nuclear magnetic resonance (NMR) signal from the nuclear spins in magnetic materials. This was supposed to be the "sure thing" experiment. Although there is no "sure thing" in science, I knew that the NMR signal in magnetic materials should be strong, strong enough, I thought, that I could detect it with a relatively simple radio-frequency bridge.

NMR in magnetic materials provides detailed information on the nature of the magnetic state. Just as Magnetic Resonance Imaging (MRI) enables one to image the inside of the body, NMR more generally serves as a non-perturbing probe into the physics that underlies the magnetism in a magnetic material. Using NMR, one can follow the temperature dependence of the magnetization, and one can study the phase transition from the ordered magnetic state to the disordered paramagnetic state. My initial goal was to use the temperature dependence of the Mn^{55} NMR signal to test the ability of spin-wave theory to predict the detailed temperature dependence of the magnetization in manganese ferrite, $MnFe_3O_4$, a material used in computer memories. Thus, I both carried out the experiments and did a thorough and detailed calculation of the temperature dependence predicted by spin wave theory. Fortunately, the agreement between the two was excellent.

In the early years of my academic life, I traveled relatively little, to a meeting of the American Physical Society once or twice a year, so I was rarely out of town. I typically came home for dinner, played with the boys for a while and then read a bedtime story to Peter. Often I read to him while we were both lying down on his bed, and often I fell asleep during the story. My falling asleep did not seem to matter much, for after one or two readings of *The Cat in the Hat* or *Green Eggs and Ham*, he knew every line anyway.

After helping to put the boys into bed, I took the car and returned to the university. Late one night in my laboratory in the spring of 1962, when I successfully detected the Mn^{55} NMR signal in manganese ferrite for first time at a frequency of approximately 650 MHz, I knew that I was on my way; I had made that first important step forward, just as I had done with the AZA youth club with the "fight" song when I started high school. I returned home that night with a smile on my face and with a light heart.

When I arrived at Penn, I had the good fortune of receiving research support from the Laboratory for Research on the Structure of Matter (LRSM). The LRSM received a block grant of funding from the Defense Department through the Advanced Research Projects Agency. A post-doctoral fellow was assigned by the LRSM director to work under my supervision, and I had sufficient funding from the LRSM to support two graduate students. As a result, my research group came quickly together.

The post-doc was from India, Dr. Sunil Ghosh — our boys soon met him and referred to him always as "Dr. Ghost." Sunil Ghosh was older and more experienced than I; he had what to me was a long list of publications compared to my paltry few. Nevertheless, I was designated as the "boss," and I rose to the occasion. I placed Ghosh in charge of the ongoing Mn^{55} NMR experiments and assigned one of the two graduate students to work with him. I put the second graduate student to work on a different but related experiment that utilized the same equipment.

I retained the higher risk project for myself. I was now confident that the NMR projects would result in nice publications. With these, and some theoretical studies that I had initiated, I was confident that I could develop, over time, a record of accomplishments that would be sufficient for promotion to Associate Professor with tenure.

The higher risk project involved an attempt to observe the electron spin resonance signal from the conduction electrons in a metal. I had a new and specific approach that I thought might enable me to succeed where others had failed. Success would have opened a new page in solid state physics. In the end, I was not successful; the "high risk" project did not live up to my hopes and expectations, but that is the nature of risk in science. The electron spin resonance in copper was successfully observed and studied 20 years later by Sheldon Schultz from the University of California at San Diego in close collaboration with Philip

Platzman from the Bell Laboratories. Schultz and Platzman had a better idea, and they developed a superior approach.

My experiments on the magnetic resonance in magnetic insulators — ferromagnetic materials, ferrimagnetic materials, and canted antiferromagnetic materials — soon began to have significant impact. I was invited to give seminars at physics departments all over the United States and invited to present my work at scientific conferences within the United States and in Europe. My "sure thing" project had turned out to be creative and worthy of attention by the physics community.

During the two years between my arrival at Penn and my promotion to tenure, I filled in some of the gaps in my education. I taught a course in solid state physics; teaching a course is the most efficient way to learn the subject matter of any topic. I listened to the lectures in the advanced graduate courses on solid state physics taught by my colleagues. Most important, I spent many hours discussing all aspects of the field with my colleagues and, in particular, in discussing aspects of the many-body physics of metals with Robert (Bob) Schrieffer.

Robert Schrieffer is a great man of science. He became famous at the tender age of 25 when he created in his mind the macroscopic, many-electron wave function that describes the remarkable phenomenon known as superconductivity. Superconductivity was discovered in 1911 in Leiden (the Netherlands) by Professor Kamerlingh Onnes. Onnes was awarded the Nobel Prize in Physics in 1913 "for his investigations of the properties of matter at low temperatures which led to the production of liquid helium." The ability to cool materials to very low temperatures that was made possible by the liquefaction of helium resulted in the discovery that some metals undergo a remarkable phase transition at temperatures near absolute zero, a transition at which the electrical resistance goes literally to zero; in a superconducting material, electrical current flows without any resistance and without any dissipation. Superconductivity, including the associated phenomena of flux expulsion and magnetic levitation (!!!), had remained a mystery and the principal unsolved problem in solid state physics — the BIG problem — for nearly 50 years until 1956 when John Bardeen, Leon Cooper and Bob Schrieffer published their BCS Theory of Superconductivity.

In the early 1960s, Schrieffer was "hot property"; he was sought after and recruited by the physics departments of many major research universities across the United States, including the physics department at Penn. A few months after we had arrived at Penn, Schrieffer and his wife Anna were invited to make their first visit to the University of Pennsylvania. Although Ruth and I were new arrivals, Eli Burstein invited us to the dinner and all the social affairs arranged as part of the recruitment of Bob and Anna Schrieffer. More important, he called Ruth and specifically asked her to join in the responsibility for entertaining and recruiting Anna. Ruth was the youngest of the faculty wives in the physics department and the closest in age to Anna Schrieffer. This well-designed and well-implemented recruiting effort was successful. Eli took Bob and Anna on the same excursion through the back roads of the Main Line suburbs to which he had introduced me a year earlier. That trip, the excitement of the opportunity to build a great physics department and, most certainly, the winning of the support of Anna, convinced Bob Schrieffer to come to Penn as the Mary Amanda Wood Professor of Physics.

I completed my education as a physicist with Schrieffer as my mentor. This was not a formal arrangement. I learned from him through many scientific discussions, through watching how he approached a problem, and through sitting in on his formal courses in the physics of metals, superconductivity and magnetism. We did not actually collaborate on a specific research project or publish together as co-authors until more than 15 years later. During the early to mid-1960s, however, I learned from Bob to aspire to achieving and creating important physics.

In order to become a great scientist, one must work with and learn from a great scientist, not just to learn the facts — the facts can be learned from books — but to learn the style and to learn what new areas of research might have high impact in condensed matter physics. I learned these in my early years at Penn from Schreiffer.

There was a remarkable group of young scientists in the physics department at Penn in the early 1960s. After the arrival of Bob and Anna Schrieffer, Douglas Scalapino and Daniel Hone came to work with Bob. Raza Tahir-Kheli, having recently completed his PhD at Oxford, came to work with Professor Herb Callen. Ruth and I became close friends with this group of young colleagues and their talented wives; a friendship that has lasted for many years.

After a period as Professor during which he made fundamentally important scientific discoveries, Donald Langenberg served as Vice Provost for Research at Penn, then moved into a senior appointment within the National Science Foundation. He was subsequently appointed President of the University of Illinois at Chicago Circle and ended his career in academic administration as Chancellor of the University System of the State of Maryland. I spoke at his retirement party and recounted the impact of his early measurements of the ratio of the electron charge, e, to Planck's constant, h.

The measurement of the ratio of the two fundamental constants, e/h, by Langenberg and colleagues started a new era in condensed matter physics. To some of our colleagues in other areas of physics, condensed matter physics had been considered as "dirty" physics; to them the complexity of real materials was daunting. One might be able to make transistors using the concepts of solid state physics, but this "dirty" physics did not measure up to the elegance of truly basic sub-nuclear physics. The demonstration by Langenberg and colleagues that the fundamental constants so precious to and revered by physicists could be measured using the phenomena of condensed matter physics changed the opinion of the physics snobs forever. Moreover, their measurements subsequently provided the basis for the new International Standard for the volt; that standard was, and remains, accepted and used throughout the world.

Raza Tahir-Kheli came from the northern hilly terrain of Pakistani-administered Kashmir. He was a brilliant student; the path from his home in Kashmir to a PhD at Oxford is a story by itself. Raza had a long and productive career as a Professor of Theoretical Physics at Temple University in Philadelphia. Raza's wife, Shirin, is the daughter of a prominent Pakistani scientist and educator, Dr. Raziuddin Siddiqi. We first met the T-Ks at a faculty party in the spring of 1962. Shirin still reminds me — and thanks me; after that party, Ruth and I gave them a ride home to their apartment in center city Philadelphia. Shirin and Raza had just married; she was and is beautiful and sophisticated, and he was and is handsome and debonair.

Shirin, too, was brilliant, and she has had a remarkable career. She somehow managed to convince her father to allow her, as a Muslim teenager, to enter university in the United States; she graduated from Ohio Wesleyan University at age 16. After her two children had started school, Shirin completed a PhD in Political Science at Penn. Her meteoric career took her quickly to Washington and to the White House, where she worked in the National Security Council with three Presidents: Ronald Reagan, George Bush and George W. Bush. She was appointed as U.S. Ambassador to the United Nations by George Bush.

Doug and Diane Scalapino, Dan and Donna Hone, Bob and Anna Schrieffer, and Ruth and I all eventually came together again in Santa Barbara. Dan Hone came first to UC Santa Barbara followed by Doug Scalapino, both in the late 1960s. Bob Schrieffer moved from Penn to UC Santa Barbara in 1980; Bob joined the physics department and served as Director of the Institute for Theoretical Physics. He remained in Santa Barbara for 10 years. In 1990, Schrieffer resigned from UCSB and became Chief Scientist of the National Magnet Laboratory at Florida State University in Tallahassee, where he also served as University Professor.

I matured as a physicist and as a scientist at the University of Pennsylvania during the 1960s. I was fortunate to be there and even more fortunate to have such a remarkable group of colleagues and close friends. I applied for my first research grant from the National Science Foundation in 1963. I was successful, although that initial grant provided only $25,000 per year. Obtaining sufficient funds to support one's research is an ongoing problem for all scientists. Over the subsequent 43 years, I have written many proposals; some were successful and others were denied. During this near half-century, however, I have been able to maintain sufficient funding to "feed" an active and relatively large research group.

Although this group of close friends moved on from the University of Pennsylvania many years ago, we remained in regular contact. For many years, we had a tradition of getting together for a New Year's Eve reunion in a different city every year. From this group emerged eight university professors, three members of the National Academy of Sciences, one Ambassador, one University President, and two Nobel Laureates. It is impossible to quantify the impact these friends and colleagues had on me, but I am certain that their brilliance, their success, and their ambition caused me to aspire to reach a level of accomplishment that ultimately resulted in the award of the Nobel Prize.

In the Spring of 1964, little more than two years after I had arrived at Penn, I was informed that I would be promoted to Associate Professor of Physics with tenure at the beginning of the 1964–1965 academic year. My future appeared both bright and secure. I had received a prestigious fellowship from the Alfred P. Sloan Foundation. The funds made available to me by the fellowship could be used to support me and my research expenses without restriction. Science is international. Thus, I chose to take a Sabbatical leave from my position as Assistant Professor at Penn to work and collaborate with colleagues in Europe, to visit laboratories in the United Kingdom and on the continent, and to experience something of the culture of the UK and Europe more generally.

There was a time long ago when people used great ships for transportation across the seas. During the early part of the 20th century, legendary ocean liners went back and forth across the Atlantic Ocean and steamships traveled between all the great ports of the world. That period is now part of history. The use of ocean liners for transportation declined in the years following the end of World War II, and dropped to nearly zero in the 1970s.

Very few people who are alive today actually traveled by luxury ocean liner to Europe and back. The Boeing 707 changed long-distance transportation; transatlantic travel is now overwhelmingly dominated by jet-engine powered airplanes. Large, luxurious ocean-going ships are still seen today, but they are used almost exclusively for cruises — for vacations — rather than for transportation from one place to another.

In May 1964, Ruth and I and our two sons were passengers on the *France*, one of the great ships of its time. We used the Sloan Foundation funds to support our first trip abroad. We were on a journey from New York to Southampton. The trip was luxurious, and it was leisurely; the transatlantic journey lasted approximately four-and-one-half days. The four of us were together in a single cabin. Although not in First Class, our cabin was quite "roomy." There were two bunk beds above one another on one wall and a single bed along the opposite wall. A small additional bed was provided, almost like a baby's bed, for David, who was only to become three years old a few months later. There were two comfortable chairs with a nice low table in front of them and a sofa-like sitting space along the wall under the portholes. Our cabin was rather high in the ship, well above the water line; we enjoyed the light from two portholes.

In New York, the ship docked at a pier on the Hudson in mid-Manhattan, somewhere in the forties. The boarding and departure were as shown so often in old films, both old films from Hollywood and old film clips taken from newsreels. There was chaotic commotion during boarding together with a great deal of fanfare and excitement. Our bags were brought onto the ship and into our cabin for us. The trunk containing the remainder of our belongings was stored in the hold below. We needed a trunk for our belongings because we were going to England for an extended period, approximately nine months, during which time I would be doing research at the British Atomic Energy Commission Laboratory at Harwell; not far from Oxford.

Science is an international enterprise. There are creative and productive scientists and great universities in Europe and Asia as well as in the United States. It is important for a young scientist to become a part of that international scientific enterprise, to learn new experimental methods and to become aware of what problems are deemed important to scientists in other parts of the world. We made new friends, and I met many new colleagues during our nine-month sabbatical. In fact, some years later when I caught the California "entrepreneurial disease" and wanted to start my first company, I was able to obtain the start-up funding from Neste, a large petrochemical company in Finland. The scientists at Neste had initiated research in conducting polymers, and they convinced the senior management to back us. But I am getting ahead of my story. This was our first trip abroad, and thus it was a particularly exciting experience for us.

The great ship departed on time on a beautiful sunny day with blue skies above. We went down the Hudson River past the Statue of Liberty and then out into the open ocean through the Verrazano Narrows. The bridge across the Verrazano Narrows was under construction; the structure was jutting out from both the Manhattan side and the Staten Island side, but there was still a huge gap in the middle. We passed far below the bridge construction on our way through the Narrows, but we could see the tiny men working on the bridge.

The *France* was indeed a luxury liner in every sense of the word. The cabins were well appointed, and the public space and dining rooms were beautiful. There was a heated swimming pool, and wide, open decks with deck chairs for relaxation. To say the least, it was not a difficult journey. The time change between New York and Southampton is five hours. When one makes the trip from New York to London today by plane, one suffers from jet-lag. On the *France*, however, the time change was made along the way in small increments; the clock was advanced by 15 minutes four times per day. One kept track of this gradual time change by looking at the clocks in the public spaces and in the cabins and by being reminded through announcements when we were informed that lunch or dinner would be served.

As expected. the food on the *France* was exceptional. This was our first encounter with French culture and in particular with French cuisine. We had, of course, enjoyed dining at French restaurants on occasion, but the cuisine on the *France* was beyond that of anything in our previous experience. The choices at each sitting were many, and they were excellent. The French bread and rolls were wonderful. A full breakfast was served, lunch comprising several courses was served, a mid-afternoon buffet was available, dinner with several courses was served, and finally, for those who were up for the late evening festivities, there was a midnight buffet. French wines and champagne were always served with lunch and dinner. The waiters took special pleasure from bringing trays of desserts around to our table and offering these delicacies to our boys, again and again. Peter and David particularly liked the chocolate éclairs, which were delicious and always plentiful.

The first day, we ate at every opportunity.

I seriously attempted to maintain a sufficient appetite to enjoy all this wonderful food. I went to the exercise room every morning to work out and made an extra effort to build up an appetite by walking the decks in the fresh ocean air, but I could not keep up with the abundance of excellent food. After the second day, we began to skip meals and did not partake of the snacks and mid-afternoon buffet. We noticed with envy and great admiration that the French people who sat at the tables in the dining room near our assigned table seemed to have remarkable stamina; they ate well and with gusto at every sitting, seemingly without difficulty, and they all looked thin!

There were day-care facilities for our boys, so we could relax, read or even sleep in the deck chairs. A deck steward would quickly bring a blanket when I sat in one of the deck chairs, for the weather was always cool on the North Atlantic. In the late afternoon and in the evening, when it was often too cool to be comfortable outside, movies were shown in the Theater. Our first trip on the *France* was a wonderful and a memorable experience.

Upon arrival in Southampton and after disembarkation from the *France*, we were pleased to find a driver waiting to take us by car to the Streatley Hills Inn, near Oxford. Formerly an English country house on the Thames, the Streatley Hills Inn had been converted to a small hotel. From the front of the hotel, we looked out across a large green lawn to a lovely view of the River Thames.

Early the next day, Ruth and I and our two boys took the bus to Oxford, where we had arranged to pick up our new car, a bright red Triumph Herald. Although "Herald" designated the model, we quickly began to refer to the car as if Herald were its given name. Herald was a convertible with a stick shift; the top could be unhooked and laid back in a moment. The upholstery was black. The dashboard was made of polished wood with the various instruments mounted into the wood. The back seat was small, not really large enough to be comfortable for adults, but perfect for our two boys. We traveled in our red Triumph convertible all over England and Scotland and three times to the Continent and back.

After taking possession of our new Triumph, our first task was to find a place to live. We moved into a small furnished house in the town of Wantage; the house at 12 Truelocks Way became our home for the nine-month period of our stay.

Because we had purchased the Triumph Herald with the intent of taking it back to Philadelphia with us, Herald had the steering wheel on the left side as in a standard American car. When driving in the UK, therefore, we had two problems. First we had to learn to drive and remain on the "wrong" side of the road. Second, the left-hand drive caused no end of difficulty. Whenever I wanted to pass a vehicle in front of us, I would start the process only to have Ruth see the oncoming traffic before I could see it. Typically that resulted in a muffled scream which caused me to pull back in line. After a little practice, staying on the "wrong" side of the road was somewhat easier. I erred rarely and only when I made a left-hand turn. During that simple maneuver, I sometimes lost my bearings and ended up facing the oncoming traffic.

We had some language difficulties with British English. We quickly learned that lorries were trucks, gasoline was petrol, aluminum was aluminium, and traffic circles were roundabouts. Some of the language problems were, however, a little more subtle.

When I returned to 12 Truelocks Way after my second day at Harwell, Ruth was pleased to inform me that because we had two young children, we were entitled to receive, free and without charge, two pints of milk every day. The milkman had come to the door earlier that day and informed her of this benefit to which we were entitled simply as a result of living in the United Kingdom. She expressed surprise and thanked the milkman. Before he turned to leave, he asked her if she wanted "regular milk" or "deer milk." Without hesitation, she responded with a smile on her face that she "would prefer regular milk." We had a good laugh over this; neither of us had ever heard of deer milk. We conjured up images of maidens chasing deer through the forest, but finally decided that in England, there must be farms on which domesticated deer were raised.

Several weeks later, I was having lunch in the cafeteria at Harwell. One of my colleagues got up from the table after finishing his food and returned to the cafeteria line for, I presumed, some dessert. He subsequently returned to our table with a small plate containing Stilton cheese and some biscuits. When he sat down he complained and said: "The Stilton looked very nice, but it was quite dear." I had something in my mouth and immediately started choking. My colleagues were alarmed and rushed to my aid. They pounded on my back until I was able to breathe again. I did not tell them about the "deer milk."

I did tell Ruth about the "dear milk." We laughed together until we cried. British English continued to retain a few surprises for us, but none were more interesting than the deer/dear milk misunderstanding.

Since I took the car to Harwell most days, we bought a used baby carriage for David so that Ruth could walk into town to drop Peter off at school and then do her daily shopping. This baby carriage was not a stroller of the kind we see today in the United States, it was an honest to goodness perambulator, wide and heavy with four big wheels. There was sufficient space behind David's seat to carry all the items that she purchased on her rounds of the various shops. There were no supermarkets in Wantage. One went to the butcher's shop, the greengrocer's shop, the hardware shop, the druggist's shop and so on.

The shopping process also caused some difficulties. When Ruth arrived at the butcher's shop for the very first time, she entered to find a number of women standing around chatting with one another; no one appeared to be waiting for an order, and no one was dealing with the butcher. Since she did not know anyone, she simply went up to the counter and said "Good Morning" to the butcher. He responded immediately and asked what she needed. This became an almost daily event. Some months later, she learned from our neighbor, Mrs. Doe, that she had broken the rules and caused some anger. The women who were standing around chatting with one another were actually in a well-defined queue; they were not formally lined up, but each person knew who preceded her and who followed her. So Ruth's unintended insult had resulted in some comments about the "rude American lady." These comments were exacerbated by the fact that Ruth typically purchased a quantity of meat nearly every day that would have been the amount purchased in a week by the other women in the queue. After she understood her error, she quickly made certain that she conformed with the local custom. She made conversation with the other women, excused herself for her error and asked for help regarding the formation of the queue. Overall, she handled this new environment with its correspondingly new customs very well.

Mrs. Doe's house was across the street from our house on Truelocks Way. She observed Ruth and her perambulator for a day or two, and then invited Ruth to join her for tea. They immediately got on well. Mrs. Doe was in her 60s, with white hair and a grandmotherly appearance. Her husband was John Doe, the ultimate nonentity. I have always referred to her as Mrs. Doe and to her husband as John; I do not know her first name. They were lovely people, hospitable and kind, and they loved our two boys. Since their own children were grown, they very much enjoyed having Peter and David in their house. At first, Ruth would simply leave two-year-old David with Mrs. Doe for an hour or two when she walked into town for shopping. Then they agreed on an arrangement whereby Mrs. Doe would care for David, and Peter would go directly to Mrs. Doe's house after school. This gave Ruth freedom to explore the area. Sometimes she would take the car, drop me off in the morning and go exploring. The city of Oxford was not far away, with many attractions. Wantage itself was interesting. She did some beautiful brass rubbings in the church at Wantage; we have them still, framed and mounted on our bedroom wall.

John Doe introduced us to King Alfred, and he introduced us to the "Blowing Stone." King Alfred lived near Wantage and ruled in the ninth century A.D.; there is a statue of King Alfred in the town square. The Blowing Stone was by the roadside and marked with a simple plaque which indicated its age and summarized its importance during the time of King Alfred. When King Alfred wanted to summon his army, he blew through the hole in the stone and made a sound that echoed over the valley. John Doe could blow through the Blowing Stone and make the ancient sound. No one else seemed able to do this. I tried but was unsuccessful. Thus, Mr. John Doe was something of a hero to Peter and David.

Near Wantage, a few miles outside of town, was the prehistoric "White Horse Hill," where the ancients had cut the outline of an abstract-looking white horse into a large area at the top of a hill. The White Horse is very large, covering an area the size of two or three football fields. John Doe also introduced us to the White Horse Hill and showed us how to get there. We visited the White Horse Hill many times; it is a mysterious and beautiful, but isolated, place. The views of the countryside from the top of the hill are quite spectacular. When we lived in Wantage in 1964, the White Horse Hill was known locally and it was noted in some guidebooks, but it was not a major tourist attraction. Often, as we walked around the site, we were the only people to be seen. When we returned for a visit in the mid-1990s, the White Horse Hill had been properly designated a national monument, and it had become a "must" stop for anyone touring England.

The next step in our relationship with Mrs. Doe was to leave the boys with her for the day when we made an excursion to London. We returned to find them very happy. For dinner, they had "beans on toast." Although they both thought "beans on toast" was an excellent choice for dinner, they were never able to convince their mother to prepare this delicacy for them.

The very first time Ruth and I were in London, I asked her if she wanted to come with me to visit "some of my friends." Although somewhat puzzled by the offer, she joined me on an excursion to Westminster Cathedral. There we saw the tombs of Isaac Newton, Michael Faraday and James Clerk Maxwell and a number of other great British scientists. I was a little surprised not to find Charles Darwin among them in that central place within the Cathedral, but he is off to one side, apparently not considered at the time of his death to be worthy of the likes of Newton and Maxwell. We then went to St. Paul's Cathedral, the remarkable structure designed and built by Sir Christopher Wren. Again, I convinced Ruth to climb the many steps to the circular gallery directly under the great dome of the cathedral in the spirit of appreciating an interesting physical phenomenon. Lord Rayleigh had named this the "whispering gallery" because sound waves were guided along the stone bench along the wall. When one even whispered, the sound could be heard clearly all the way across the other side of the dome.

On weekends, we put the boys in the backseat of the Herald and took off to see something of the country. We went to Stonehenge. At that time, there were no boundaries and no fences; one could wander among the huge and ancient stones without any restrictions. We visited most of the better-known castles in Devon, in Cornwall and in Wales, and we visited many of the great cathedrals. Typically, we would depart on a planned weekend journey on Saturday morning, spend Saturday night at a bed and breakfast (often a private residence with a small sign out front) at our destination, and return to Wantage on Sunday afternoon.

Ruth and I traveled to the Continent three times in our Herald. The first time we went alone, leaving both Peter and David in Mrs. Doe's care. We departed from Wantage in the morning, drove through London and out toward the English Channel, and stopped for the night in Dover. We found a restaurant in Dover to which we returned several times on the way to and on the way back from the Continent. At this restaurant, they served Dover sole followed by a wonderful Irish coffee that served to warm me and to give me a good night's sleep. Of course we explored the White Cliffs of Dover while singing the song made famous during the Second World War.

The ferry ride across the Channel was always a harrowing experience. The challenge was to attempt to make the crossing without getting seasick. My approach was to sit in a chair on the middle level deck and to bury myself in a book. I had to be careful not to look up, because if I looked up from my book, I would see vomit sliding down the windows from seasick passengers on the upper deck. These unfortunates were unsuccessfully trying to recover in the fresh air from the sea. One glance at that sight, and I would get sick as well. Sometimes we made it, sometimes we did not.

During our three trips to the Continent, we visited major cities in Belgium, Germany, France, the Netherlands, Denmark and Sweden. In each city, I made arrangements to visit the physics department at the university. I discussed my work with colleagues there, and listened to summaries of their research. Usually I gave a lecture. The highlights were Amsterdam, Berlin, Copenhagen, Stockholm and Gothenburg and, of course, Paris. In Berlin, we also enjoyed the nightlife and an excursion by tour bus through Checkpoint Charlie into Russian-occupied East Berlin. In Amsterdam, we rented bicycles and did our sight-seeing, including a visit to the Anne Frank House, from the seats of the bicycles. In Copenhagen our rented bicycles took us to visit the Little Mermaid, and they carried us to many furniture stores, where we purchased Danish modern furniture and arranged to have it shipped to our home in the United States. We enjoyed boat tours through the canals of Amsterdam and again through the waterways of Stockholm. In Stockholm, we went to the Opera to experience *The Barber of Seville* sung in Swedish.

We were hosted in Gothenburg by Professor Stig Lundquist. We knew Stig because a year earlier he had taken a sabbatical leave from Chalmers University in Gothenburg, where he was Professor of Physics, and spent the period of his sabbatical as a Visiting Professor at Penn. Stig showed us the city and its surroundings, he hosted us at dinner, he introduced us to aquavit, and he took us to a concert in the wonderful concert hall in Gothenberg.

Stig was a world-class theoretical physicist with a strong international reputation. He subsequently became a powerful and influential member of the Nobel Committee for the Physics Prize and served as Chairman of the Physics Committee for a dozen years. He played a very important role in the lives of the physicists who became Nobel Laureates during those years.

In 1983, not long after I had moved to Santa Barbara, Stig came for a visit and remained as Visiting Professor in the physics department at UC Santa Barbara for a few months. While he was in Santa Barbara, I was informed that I had received the Oliver E. Buckley Prize. The citation read as follows: *"For his studies of conducting polymers and organic solids and in particular for his leadership in our understanding of the properties of quasi-one-dimensional conductors."* The Buckley Prize is an important statement of recognition; the Prize is awarded every year by the American Physical Society for significant scientific accomplishments in the field of solid state physics. The Buckley Prize was the first such statement of recognition awarded to me. When I learned of my good fortune, we quickly organized and announced an open-house celebration party for that very evening. Many friends and colleagues from the university came to our home to congratulate me and to celebrate with me with a glass of champagne. During the evening, Stig rose and made a toast in which he congratulated me on this award. Then he said: "In Sweden, we are watching your progress." Evidently, as we learned 17 years later, they were indeed watching.

On our second trip to the Continent, we took Peter with us. Again, we started in Bruges, toured Amsterdam and the Netherlands, where we discovered the Kröller-Müller Museum. At that time, the museum was housed in a relatively small farmhouse within a large nature reserve that subsequently became a National Park. The Kröller-Müller Museum has since been rebuilt; the Van Gogh paintings are now properly shown in a well-designed and well-lighted modern building. The paintings, however, are the same. We have returned several times to see this unique collection, a feast of Van Gogh's paintings.

We went to Paris and found a small hotel on the Left Bank. An early destination was, of course, the Louvre. We sailed boats in the Tuileries, we ate pâté and cheese on baguettes, we went to the top of the Eiffel Tower, and we explored Sacré-Coeur and Montmartre. We generally took our dinners at good restaurants — not restaurants with stars from the Michelin guide, but good local restaurants recommended by the hotels where we stayed or by local shopkeepers with whom we had talked. Our son, Peter, was the "star" at these evening dinners. He was an engaging little boy. He was willing and eager to try new foods, even dishes as exotic as escargots, and the waiters always paid attention to him. We were treated much better in France when we traveled with our son than when we traveled alone.

After leaving Paris, we went south to Provence. We saw our first Roman ruins and explored the ancient sites. We stood on the old bridge in Avignon and sang together the French children's song, "Sur le Pont d'Avignon."

On the way back home we managed to get through Paris, across the English Channel, through London and back to Wantage all in one long day! David was pleased to see us when we returned, but not at all anxious or concerned. He had a terrific time playing with his Matchbox cars while staying with Mrs. Doe and eating "beans on toast." The only mishap had occurred when a bee landed on the sidewalk in front of him, and he decided to run the bee over with his small model of a double-decker London Bus. Mrs. Doe soon quieted the tears, fixed the hurt finger and put a small bandage in place, so that when we arrived he was just fine. Nevertheless, he described the incident to us in great detail.

Although we traveled extensively, we also enjoyed the English countryside near Oxford. We often had tea at the White Swan Hotel on the Thames at Abington; Peter and David went out by the locks and watched the boats while Ruth and I relaxed and looked on. A boat that came downriver went into a lock, sat quietly as the gates were closed, dropped as the water was let out, and then proceeded further downriver. Because the big water-gates were operated by hand, the process was necessarily slow and leisurely. Enjoying tea at the White Swan Hotel and the nearby activities on the river for hours on a Sunday afternoon or on a summer evening remains in my mind as a special pleasure.

During all this traveling, drinking tea and exploring, I managed to get some work done. My goal was to complete a series of magnetic susceptibility measurements on a set of samples I had prepared in my lab at Penn and brought with me. The technician at Harwell, John Penfold, was a no-nonsense guy; he was careful and obtained high quality data. I had little to do — he would not allow me to touch his beloved susceptometer — so I needed only to accumulate the results and to think about their implications. At the end of our stay in England, my briefcase was filled with data. When we returned to the United States, these results provided my research group with a jump start into a new direction.

Ruth and I made one last trip to France before returning to the United States. We went directly through Paris and headed south, first to Lyon and then on to Marseilles, the French Riviera and Monte Carlo. The French Riviera was a fitting way to end our first experience in Europe. We had one problem which put a damper on our enthusiasm. On the way back to England, on a narrow road in the south of France, we had an automobile accident. The oncoming car was going too fast for the curve on the road and skidded into the Triumph, damaging the door on the driver's side.

When we returned to England, we said goodbye to Mrs. Doe and John and wrapped up our affairs in Wantage. I dropped the damaged car at the dealer in Oxford and arranged to have the door repaired before it was placed on a ship to be sent to the Port of Philadelphia. When Herald arrived in the United States, he was once again whole and pristine.

I loved that little Herald. After returning to the Philadelphia area, it gradually decayed, principally as a result of the corrosion from the salt poured onto the Philadelphia streets following the winter snowfalls. I nursed the Triumph along for several years beyond its natural life. Finally, in 1974, after the floor rusted through so that I could see the street below and after the clamps that held the battery in place rusted through so that the battery fell out when I stopped at a traffic light, I decided that the time was approaching when I must say goodbye to Herald. The final parting came, however, when the left front wheel came off as I was commuting to the University of Pennsylvania on the Schuylkill Expressway at relatively high speed in the midst of the morning rush hour traffic. The car lurched and screeched to a halt — luckily without hitting or being hit by another car — and the perfectly balanced left front wheel rolled on and on and on and on, remaining perfectly centered in the inside lane of the expressway. I jumped out of the car and ran as fast as my legs would carry me in pursuit. Eventually, I retrieved the wheel, rolled it back to the waiting car, jacked-up the left front end, removed one bolt from each of the other three wheels, re-mounted the errant wheel, took down the jack, and — I was on my way! I accomplished this without anyone honking or yelling vile curses at me for causing them to be stopped and delayed on the expressway. Evidently the sequence of actions and activities was sufficiently amusing to keep everyone but me entertained. The following weekend, I traded Herald in for a new Volkswagen Beetle.

Although the return trip from London to New York was by plane, we went again, two more times, by ship across the Atlantic Ocean. We traveled on the *Michelangelo* from New York to Genoa in 1968 and on the *Nieuw Amsterdam* from Rotterdam to New York a year later. These were our last real journeys by ship; as I noted earlier, the use of ocean liners for transportation declined in the years following the end of World War II, and dropped to nearly zero in the 1970s.

The period in England, in Wantage, and our excursions on the Continent during this period had significant impact on our lives. We departed from New York on the *France*, unsophisticated, inexperienced and even somewhat fearful of being so far from our home and our roots. We returned forever changed as a result of this first rewarding and enriching travel experience.

During the past 30 years, I have traveled all over the world, many times to various countries in Europe and Asia, to South America, to Australia and to New Zealand. As a family we spent a sabbatical year in Geneva (1968–1969) and again traveled widely throughout Europe. In more recent years, my trips have become tightly scheduled for the purpose of attending and lecturing at science conferences or sometimes for the purpose of promoting my entrepreneurial activities. I have accumulated more than 3,000,000 miles on United Airlines alone. When such a trip offers a chance to spend a little quality time in one of our favorite places, Ruth accompanies me. We have indeed become world travelers.

10 "Fire and Rain"

"Fire and Rain" is the name of a ballad written and sung by James Taylor. When I hear this song, either from a CD recording or a "recording" in the memory bank of my mind, I think of my friend Tony Jensen. In the first verse, Taylor sings:

> I've seen fire and I've seen rain.
> I've seen sunny days that I thought would never end,
> I've seen lonely times when I could not find a friend,
> But I always thought that I'd see you one more time again.

As in the lyrics of this song, I always thought that I would see Tony "one more time again", but he died a long time ago.

Among experimental physicists — that is, physicists who do experiments to investigate Nature's secrets in contrast to theoreticians, who try to uncover Nature's secrets through intuition, insight and creative mathematical analysis — there is a tradition of carrying out experiments late into the night. Late at night, there are fewer people around, less electrical noise from machinery and electronic equipment, and reduced vibrations from traffic outside or movement inside the building. Any of these can interfere with the sometimes weak signals that must typically be detected in any forefront experiment. This is especially true in the field of low-temperature physics, where experiments are carried out on samples that have been cooled to ultra-low temperatures, temperatures approaching absolute zero (minus 273°C).

During the period from 1966–1968, Tony Jensen and I spent many late nights in the Laboratory for Research on the Structure of Matter on 33rd and Walnut Streets at the University of Pennsylvania, working together on low-temperature physics experiments and attempting to create a theoretical solution to a difficult physics problem.

Tony was a like a meteor, "a shooting star" in my life; he appeared in the autumn of 1966 when he joined the physics department at Penn. He had a powerful intellect and a deep interest — perhaps better described as a love — of science. He was one of those rare people who seem to know so much more than they have had time to learn. He taught me a great deal about science and a great deal about myself. His light burned brightly, sometimes too brightly, and then five years later, he was gone. After his death in 1971, I missed him, and I thought of him nearly every day. For a long while, I held on to the guilt that originated from the feeling that I had let him down and not been there when he most needed me. But time heals, and now I hold onto only the good memories.

Tony left a mark on everyone who knew him. He exuded an intensity and a love of life that were infectious; he got into one's soul in a way that I had never previously (nor since) experienced. Although the memories and emotions have diminished in intensity over the 40 years that have passed since he died, I still think of him now and then. When I do, the memories are pleasant, sometimes even funny, and cause me to smile.

When I returned from England early in 1965, I shifted my research into a new direction; I focused on the subject of localized magnetic moments in metals. In particular, how did the localized spins of magnetic impurities in metals interact with the conduction electrons, the free electrons, of the metal? This problem had received a great deal of attention following the discovery in 1964 by a Japanese scientist, Jun Kondo, that in the presence of dilute quantities of magnetic impurities such as Iron (Fe), the electrical resistance of metals *increased* as the temperature was lowered. Kondo provided an initial theoretical analysis which predicted an increase proportional to the logarithm of the temperature $(-\ln T)$, in agreement with the experimental results that had been previously obtained in a number of laboratories in the United States, in

Europe and in Japan. Because of his insight, this increase in resistance as the temperature was lowered was known as the Kondo effect. The Kondo effect was unusual and anomalous; the electrical resistance of metals typically *decreases* as the temperature is lowered (because the atomic vibrations decrease as the temperature is lowered) and finally, at sufficiently low temperatures, becomes independent of temperature. Since ($-\ln T$) diverges toward infinity as the temperature approaches zero, Kondo's logarithmic temperature dependence could be only a first approximation. Often in physics, such an apparently divergent term signals a transition to a new state. Thus, the fundamental question focused on the nature of the lowest energy state of the magnetic impurity problem or the "ground state of the Kondo problem."

During the period from 1965–1969, the Kondo effect was considered to be one of the important problems in condensed matter physics. The Kondo effect was, however, a high risk problem. There had been no significant progress in theoretical understanding since Kondo's stimulating publication. Although the resistance measurements were relatively straightforward, the experiments that were most needed were those that could somehow provide information on the correlations between the electron spins on the magnetic impurity and the spins of the conduction electrons in the metal. Experiments designed to probe the spin correlations at low temperatures were complex and certainly not straightforward.

Jensen had completed his PhD thesis research in physics at the San Diego campus of the University of California. His thesis research was on aspects of superconductivity; his work focused on attempts to increase the transition temperature to the superconducting state by carrying out systematic measurements on a variety of alloys and inter-metallic compounds. When Tony arrived at Penn, he initially continued to focus on superconductivity, but he was of course aware of, and interested in, the magnetic impurity problem.

Tony and I began to collaborate on research directed toward a deeper understanding of the Kondo effect; a collaboration that developed into a close friendship. He had experience in the methods of producing ultra-low temperatures, and I had experience in carrying out magnetic resonance experiments. We worked together on a series of nuclear magnetic resonance experiments on metallic alloys containing dilute quantities of magnetic impurities (at concentrations of a few parts per thousand or even less) at very low temperatures; for example, dilute alloys of iron in copper or manganese in copper. Our goal was to probe the nature of the spin polarization around the magnetic impurity as a means of learning about the ground state of the Kondo problem. With the powerful information that could be obtained from these experiments, we were convinced that we would be able to lead the way — lead our theoretician colleagues — toward the solution. Our collaboration was intense, it was productive, and it was fun.

There we were at in the middle of the night, in the lab with the "demagnetization" apparatus in operation and with our samples at low temperature. Tony speculated to me that "at that precise moment, our experiments were being carried out at the lowest temperatures of any place in the world." After one particular all night marathon of accumulating data, I came home and entered the front door in the morning just as Peter was coming out of his room to get ready to go to school.

In our family, we have never referred to what I do as "work." I go to "school," or to the university. I am a professor, I teach physics and I do research, but never do I *work*. Thus, when he came out of his room and saw me coming in the door, he must have thought to himself that I was coming home from the university or from "school." After going to medical school and completing his MD and residency, Peter became a research immunologist. The "research genes" are evidently dominant genes. When I call his home and one of his two sons or his wife answers the telephone, they never say to me that he is at work; they tell me he is "in the lab." Similarly, when I call my other son, David, who is a professor and a neuroscientist, I am often told that he is "at the university." We are very fortunate in our family: No one works.

Tony and I also collaborated on theory as well. We attempted to solve the problem of the ground state of the Kondo Problem; that is to formulate the many-electron wave function that would describe the lowest energy state and predict its properties. This, too, was the subject of many long nights of effort in our offices and in the empty hallways of the LRSM. We understood that the ground state must be of singlet character with the spin of the magnetic impurity fully compensated by the build-up of spin correlations within the electrons of the metal. The question was how to describe that many-electron singlet ground state. After much discussion, we focused on a specific approach that was motivated by Schrieffer's famous wave function which described the ground state of superconducting metals.

We spent one of our long nights in the LRSM formulating that idea in mathematical terms, evaluating the ground state energy and then using the minimum energy principle to determine the functional form of the unknown parameters in the wave function. I worked in my office on part of this calculation; Tony was doing the same on a different part of the calculation in his office on the other side of the building. We stopped to compare notes periodically, and got more and more excited as the night wore on. "I am certain we are on the right track. This just feels right," said Tony at one point.

We combined our efforts and completed the final steps of the calculation on the blackboard in my office. In the middle of the night, everything seemed to come together. We had created a many-electron singlet wave function that satisfied the minimum energy principle. It looked right; we thought we had the answer! We were so high, so full of joy at the success of our effort that we decided to go out and celebrate. Moreover, we were hungry; we had skipped dinner in the pursuit of science.

In Philadelphia, the cheesesteak sandwich is famous. The cheesesteak is not fast food, it is a sandwich prepared to order; thinly sliced meat cooked with onions on a grill and served on an Italian roll with something like Cheez Whiz poured over the meat and onions. If one ignores the high cholesterol content, a good cheesesteak is a very special sandwich. In Philadelphia, there were many places to go for a cheesesteak, but only a few offered the genuine product. Of these, only a very few are "original"; Pat's Steaks is one of those places where one can find an "original" cheesesteak. In the late 1960s, and I presume even today, Pat's Steaks was open, serving delicious cheesesteaks all night every night. There is no restaurant to go inside of; one goes up to a window, orders and pays, and then receives the order from the next window. All kinds of people show up at Pat's in the middle of the night; workers coming home after their shift, couples dressed in gowns and tuxedos after a night out, and sometimes even professors from the university who have interrupted their research in search of a cheesesteak.

The two professors took off from the university toward South Philadelphia in Tony's two-seater sports car, to Pat's Steaks. We had not had anything to drink; we were cold sober but in a deliriously happy mood. The cheesesteaks lived up to our expectations. On the way back to the university, we raced up Walnut Street at speeds in excess of 80 miles per hour while gleefully agreeing that if we crashed and were killed, no one would ever know what we had accomplished. Luckily, we lived through the wild ride and there were no police in sight at that early hour.

We had not solved the Kondo Problem. Although the results from the Jensen–Heeger collaboration, both experimental and theoretical, did contribute to the field and provide insight into the fundamental unresolved issues under study, the problem turned out to be remarkably difficult and resisted the efforts of the best theoretical physicists in the world of that period.

The Kondo Problem was finally solved by Kenneth Wilson in 1974 using the numerical renormalization methods that he developed for solving the much broader problem of phase transitions in solids; for example, the transition from ferromagnetism to paramagnetism. Wilson was awarded the Nobel Prize in Physics in 1982 for creating a general theory of phase transitions, a theoretical structure that predicted all aspects of the data on phase transitions that had been accumulated by physicists over many years.

The Wilson solution did not open up a new field of science as did, for example, the Watson–Crick discovery of the structure of DNA. The Kondo effect appeared to be interesting (it is interesting!), but the solution did not lead to more general insights into other problems, nor did it provide a foundation for a new technology. Nevertheless, physicists continue to study materials that are in the so-called Kondo ground state.

One only knows if he/she has chosen a good problem after it has been solved, or after a breakthrough has occurred. Nevertheless, as a scientist, I continually think about my choice of problems, and I try to be ruthless in dropping a high risk project when it becomes more or less clear that the impact will not justify my effort.

I remember a conversation with my son David on this topic. He had finished his PhD in Computer Science and was about to take a post-doctoral appointment in the Media Lab at MIT. We were taking a long walk and he expressed concern over this very point. I responded that one can only do his "job" and do it to the best of his ability. If you do that, with a little good luck, you might make important contributions to science. At the Media Lab, he shifted his interest toward psychology and then to neuroscience. He is now a Chaired Professor of Neuroscience at New York University. He did his job, again and again, and he did make important contributions to psychology and neuroscience. He was inducted into the National Academy of Science in 2014. Obviously Ruth and I were very proud to witness his induction at the NAS Meeting in Washington. We are not the only father-and-son pair in the National Academy of Science (scientists tend to breed scientists), but there are relatively few such father–son pairs.

My older son Peter is a medical doctor and Chaired Professor of Immunology at Mount Sinai Medical School. He does research on the problem of transplant rejection. He has no such difficulties with the choice of where to focus his work. The importance of the problem associated with preventing rejection of transplants in human patients is absolutely clear.

But I have digressed and must return to my love affair with Tony Jensen. We were in Washington, D.C. in March 1968 attending and participating in a meeting of the American Physical Society. Since this APS meeting took place during the period of our collaboration on the Kondo effect, we had new results, and new ideas based upon our results, to describe and to discuss with our physicist colleagues. At the end of the second day of the meeting, we went out to dinner with a group of friends and acquaintances, a leisurely dinner accompanied by wine and by heavy scientific discussion. After finishing and appropriately splitting the cost of the dinner, Tony tried to convince the others from the dinner group that we should go out on the town. I had already had enough; I was tired and indicated that I was going to head back to the hotel and go to bed.

Before we parted from our dinner, Tony took me aside and told me that later he would likely bring a woman back to the room we were sharing at the conference hotel. With a smile on his face, he said that when he did, I was to get out of the room without delay and leave him with some privacy. I laughed and expressed doubt about his boast, but promised that if indeed he were successful, I would quickly get out of his way and out of the room.

Friends who knew Tony when he was a graduate student at UC San Diego have told me that he was serious, focused and happily married. Sometime toward the end of the period of his thesis research, he "looked up" and took on other interests. He was handsome and attractive to women. He had intense blue eyes. One look from those blue eyes seemed to have an amazing impact on any woman. By the time he arrived at Penn, although still married, he was casting about with interest in women in general and with specific interest in a select few. Thus, his plan to bring a woman back to the hotel room in Washington did not come as a complete surprise to me.

Hours later, sometime during the night, I was awakened from a deep sound sleep; Tony shook me awake and urged me to move quickly and get out of the room. Although not yet really awake, I found myself being guided out the door. The door closed and there I was all alone in the hallway. By now I was awake and began to understand my predicament. I had nowhere to go; I was in my pajamas and becoming seriously angry. I was about to pound on the door to "my" room and demand that I be let in, when the door opened and Tony fell into the hallway laughing. There was no woman; he had simply set me up.

I was a physics professor at an Ivy League university and that meant that I should have certain tastes. For example, listening to and dancing to rock and roll music and listening to the Beatles were outside of my domain of interest. Tony taught me to loosen up and avoid the stereotypical images. Early in the development of our relationship, he bought for me the Beatles' record *Sgt. Pepper's Lonely Hearts Club Band.* He gave me the record, and he sat down with me and insisted that I really listen. Within a few minutes, I understood and, for the first time, appreciated the talent and creativity of the Beatles.

A bar with a rock band and dancing had recently opened at the Marriott Hotel on City Line Avenue, not too far from where we lived in the Main Line suburbs. Tony coaxed us out onto the dance floor and showed us the way. After that introduction, Ruth and I went regularly after a movie or after dinner to dance, enjoy the music and enjoy ourselves, sometimes with Tony, sometimes alone. We bought records and invited our university friends and my physicist colleagues to our house for "wild" parties with rock music and dancing.

We had just purchased a new car, a four-door "family car" that was appropriate to a professor with a wife and two young children. I described the new car to Tony with pride, and I took him for a ride. He verbally approved, but his eyes told me: "No — this is not fun — you only live once, so enjoy your life!" Today, I drive a two-seater Mercedes sports car, a hard-top convertible. With the push of a button, the top drops back into the trunk in just 20 seconds. Each time I put the top down, Tony Jensen flashes through my mind.

He transformed me in other ways as well. We discussed the Vietnam War many times and at great length. Because he made me face the real issues, I emerged as a serious opponent of the war.

Tony was not satisfied with words; he was an activist. When the revolution in Czechoslovakia broke out in 1968, he went to Prague. He was caught by the Russian police while carrying and using a tiny camera that fit in the palm of his hand. He was taken into custody. Although he was released after 24 hours, the Russians forced him to leave the country. When he returned to tell me the story in detail, he claimed to have had a love affair in Prague during the short period before the Russians caught and detained him. He probably did.

I have never been more open with anyone. We were able to talk about anything, even the most personal parts of our lives — about our hopes, about our insecurities and about our fears. He was my friend; he was the second great love of my life. I will never again be able to allow myself to be so close to another person other than Ruth; the loss was too painful.

In the summer of 1968, Ruth and I and our sons departed for Geneva where we spent my sabbatical year. It was a great year, a wonderful year, and I described in detail earlier in this narrative how learning to ski in Switzerland contributed so much to the enrichment of our lives.

When we returned a year later, there was a clear change in Tony. He seemed to no longer be interested in science. Rather than continuing his research on superconductivity, the Kondo effect or the insulator-to-metal transition, he had decided that he would move in new directions. He had requested that the staff of the physics department machine shop build a large light-tight wooden box and to line the box with sound absorbing foam. He was planning to sit quietly in the box for long periods of time in order to experience the stress of sensory deprivation. The analogy with a tomb was to me quite obvious.

He and his wife had divorced. She had remarried and moved with her children back to the Los Angeles area. Tony was not living alone, but he had not made any long-term commitments. He had made certain that there were "no strings" attached.

I did not (and do not) consider Jensen to be manic-depressive. He was not "crazy" in the usual sense of the word. He had a fear of mental instability — he had expressed that openly to me — principally because his mother had a history of problems, but his father, Melvin Jensen, was a healthy and successful businessman.

The clues to understanding his suicide must be in the facts that he had separated himself from his family and that he had stopped doing science. He was not able to handle the intensity of life as he lived it, and he could not live his life any other way. He could no longer cope with the risk of long-term commitment, he could no longer cope with the inherent risks of science. He had lost his nerve because he was unable to cope with the biggest risk of all for a creative person, the risk of losing his creativity as he grew older. He jumped from the top of a 33-storey building. I had seen and talked with him at some length only two weeks earlier. I did not anticipate his suicide. Perhaps I simply did not pay close enough attention.

During the month of July 1971, Ruth and I, with Peter and David, were living in Aspen. During the decade from 1965 to 1975, we spent many summer months at the Physics Institute in Aspen. If you ask my sons about their childhood, you will quickly find that their summers in Aspen have an important place in each of their memories. The Institute arranged housing for the participating physicists, provided bicycles for their families and arranged regular excursions and hikes into the mountains. The Aspen Music Festival provided endless opportunities for enjoying outstanding music played by world-class performers. Then, as now in Aspen, there were outstanding restaurants from which to choose. But the mountains, those spectacular Rocky Mountains, were the main attraction.

The Physics Institute provided a perfect environment for writing up my research results for publication, for planning the next steps in my research, and for writing proposals for future funding. There were always outstanding scientists in residence, and interactions among participants were both natural and encouraged. Because of these facile interactions, I made important progress on a number of problems during my summers in Aspen.

I received the call from Philadelphia, from the police department on July 7, 1971, shortly after I had returned to our condominium from the Physics Institute. The police had been informed that I was Tony's closest friend, and that I would know how to contact his family.

I called Melvin Jensen. I told him that Tony was dead, I told him what little I knew of what had actually happened, I told him how sorry I was to have to be the bearer of such bad news, and I told him that the police in Philadelphia needed to have him, as Tony's closest relative, identify the body. We did not talk for long — there was little to say. Mr. Jensen called back later that day; he had begun to make the funeral arrangements. Tony would be buried in a cemetery in Los Angeles after a service in the Mormon Temple in West LA. He asked me to say a eulogy for Tony at the funeral service.

Immediately afterward, I made contact with friends and colleagues at Penn. A memorial service was planned, to be held two days later at the university. I was asked to say a few words about Tony at this service. Flight reservations were quickly made; first from Aspen to Denver and then from Denver to Philadelphia. I would spend two days in Philadelphia and leave shortly after the conclusion of the memorial service for the airport and a flight to Los Angeles. Finally, I reserved a flight from LAX to Denver and then to Aspen to rejoin my family, with departure from LAX scheduled early in the morning following the day of the LA funeral service.

On the flight from Denver to Philadelphia, I wrote the comments that I would make at the two services. Although the "audiences" at the two services would be very different, I focused my comments on Tony as a person, on his remarkable talent and creativity as a scientist, and on his scientific accomplishments. I hoped that these comments would be meaningful to both groups.

The night before the memorial service at Penn, I spent several hours with two of Tony's friends, close friends of mine as well, Paul Chaikin and Joel Cohen. Paul was a graduate student working on his thesis under Tony's supervision; he was mature beyond his years — both as a person and as a scientist — and more like a colleague than a student. Joel held a post-doctoral research position in the physics department. Both were close to Tony. We drank quite a lot of Scotch whisky, we talked a lot, we listened to Tony's favorite music, and we began to try to confront the reality of losing him.

We spent quite a long time discussing what music would be played the next day as people assembled and sat quietly at the beginning of the service; the music became of prime importance to us. Paul agreed to assemble our choices on a CD in a specific order, and he agreed to arrange to have the necessary equipment in the room where the memorial service was scheduled. I do not remember all the pieces that we played. The two that stand out in my memory are "Fire and Rain" and "You've Got a Friend"; both are ballads with relatively simple melodies by James Taylor. The lyrics were particularly meaningful to us that night. "Fire and Rain" is another of the many melodies that continue to pop into my head at random times.

After the memorial service, I quickly said my goodbyes to friends and colleagues, caught a taxi and went directly to the airport. I had more time, but did not want to share the time; I simply wanted to be alone.

The flight from Philadelphia to Los Angeles seemed endless. I had "buried" Tony once, but now I had to do it again. The service in LA was far more difficult for me, in part because it was the second time that I gave the eulogy, in part because the surroundings were unfamiliar, and in part because I was there without the support of people that I knew.

I returned to Aspen four days after I had departed, exhausted and depressed, just before noon on Saturday. Ruth and the boys met me at the airport; it was another beautiful day in Aspen. After stopping at our condominium to leave my bag and change clothes, we headed up the road toward Independence Pass to a favorite spot that we had visited many times before.

We took off our shoes and waded along a relatively flat stretch in the Roaring Fork River, before it cascaded down toward the valley below and passed through Aspen. We then went farther up to the "Devil's Punch Bowl," an area where the river flowed steeply among huge boulders between a series of deep pools. There were young people, mostly young women, sunning themselves on each of these boulders, naked and beautiful. This remarkable vision restored me. I took off my clothes and jumped into the cold water, with Peter and David immediately following me.

Tony would have approved, and he would have been proud of us.

Tony's death had a lasting impact on me. His inability to deal with the eventual loss of creativity gave me both the strength and determination to prove that I could continue to be productive and creative as I grew older. Today, at age 79, I continue to be a world leader in my field and, remarkably, I have demonstrated some important creative insights into several problems within the 79th year of my life.

11 Risk, Creativity and Discovery – Again

The degree of risk involved in an endeavor cannot be quantified in absolute terms. The degree of risk depends on the talent, the knowledge base and the skill of the person who is about to take on that endeavor. My son David is a very good skier; he skis at a level well beyond what is commonly termed Expert. David can safely go down steep and narrow slopes with ease and grace; he attacks such terrain without taking significant risk. Although I, too, call myself an Expert skier, were I to follow him down these most difficult ski runs, I would be taking great risk, risk of falling and severe injury or even death. In sports and in science, the degree of risk depends on the talent, the knowledge base and the skill of the person who is about to take the risk.

By 1975, I had matured as a scientist and significantly expanded my core scientific knowledge base. Because of my experiences during the TTF-TCNQ saga, because of the broad range of skills that I had to learn as a scientist who does experiments on "new materials" — materials that are synthesized by chemists rather than materials that are found in Nature — and because of my deeper understanding of the possibility of uncorrelated events as in the story of the "penny and the wart," I was well prepared to take on a relatively high risk scientific problem. My experiences during the early 1970s had given me the confidence, and the strength of character, to have courage in a time of scientific uncertainty and resulting controversy. Those experiences had also taught me that in the world of competitive creative people, one must have the intellectual strength and scientific power to fight the "lions," the "thieves" and the "enemies who come to dispute one's claims."

In 1973, a particularly interesting paper appeared in the prestigious journal, *Physical Review Letters*. M. Labes and colleagues in the chemistry department at Temple University introduced the inorganic polymer known as poly(sulfur-nitride), or $(SN)_x$, where S is the chemical symbol for sulfur, N is the chemical symbol for nitrogen, and the subscript "x" is intended to denote the number of sulfer-nitrogen repeat units in the polymer. If the number of repeat units was known and established by one of the well-known methods of polymer science, poly(sulfur-nitride) would be referred to as $(SN)_n$ where n, the so-called Staudinger index (after Hermann Staudinger, the scientist who first understood the structure of linear polymers), is the number of repeat units. When the number of repeat units is unknown because, for example, the polymer is completely insoluble in any solvent so that the methods developed by polymer scientists are not applicable, the Staudinger index is appropriately replaced by "x." Labes and colleagues showed in their first publication that poly(sulfur-nitride) had the electrical properties of a metal; $(SN)_x$ is an excellent conductor of electricity and even looks like a metal, shiny and gold colored.

Because this was the first example of a metallic polymer (although inorganic), I lusted for the opportunity to initiate research on this novel material.

After making a few inquiries, I learned that a professor in the chemistry department at Penn, Alan MacDiarmid, had experience in sulfur-nitrogen chemistry. MacDiarmid was born and grew up in a small town in New Zealand, near Wellington. He completed his MSc in Chemistry at Victoria University in Wellington before coming across the world to do PhD studies at Cambridge (UK) and subsequently at the University of Wisconsin. MacDiarmid had done his Master's thesis on sulfur-nitrogen chemistry.

I was acquainted with MacDiarmid, but I did not know him at all well. He was an inorganic synthetic chemist with scientific interests far removed from my own. MacDiarmid's research focused on silicon chemistry, that is on the synthesis and study of various molecules in which the element, silicon, played a central role. He was the leader of a relatively large research group consisting of graduate students working toward their PhD degrees and post-doctoral researchers.

As Director of the Laboratory for Research on the Structure of Matter, I had responsibilities for evaluating the research of others in the context of their ongoing participation in the LRSM scientific program. Primarily in this role, rather than because of any personal interest in his work, I had heard MacDiarmid give a few lectures at research reviews. He always started his lectures with a comment that silicon was one of the most common elements on our planet, and he used this fact as partial justification for his focus on silicon chemistry. He was also fond of reminding the audience that on the periodic table of the elements, silicon resides directly below carbon. All of organic chemistry and biochemistry is based upon carbon. MacDiarmid planned, as his life's work, to explore the analogous chemistry of silicon.

My point in briefly (and incompletely) summarizing MacDiarmid's research prior to 1975 is simply to make it clear that he was an active and successful chemist with a long-term commitment to a specific scientific area. Nevertheless, because I found it remarkable that poly(sulfur nitride), $(SN)_x$ was a synthetic metal — a metal that contained no elements from the periodic table that are known to be metallic — and because I lusted for the opportunity to work on this new metallic polymer, I was certain that I could convince him to change his scientific focus and collaborate with me. Although I succeeded, the hoped for collaboration almost did not happen because of a classic misunderstanding.

I contacted MacDiarmid by telephone, briefly told him of my interest in this new synthetic metal and made an appointment to meet with him in his office to discuss the opportunity in greater detail. Our meeting started relatively late in the afternoon of a beautiful, sunny, autumn day in 1974. MacDiarmid's office looked out to the west through large windows. When I arrived, the sun was streaming in and onto the floor of the room. We sat together, on the opposite side of the room from the windows, around a small table in front of his desk and near a blackboard that was mounted on the wall.

After a few minutes of "chit-chat," I brought up the subject of $(SN)_x$ and my interest in doing experiments on this new material. He was polite, but did not seem particularly interested. I went into a selling mode and attempted to describe to him that the metallic nature of $(SN)_x$ was of fundamental scientific interest and a special opportunity. This somewhat one-sided conversation went on for quite a long time, roughly an hour, during which the sun had dropped lower in the sky so that the direct sunlight from the windows was now directly on us and on the small round table. Finally, with a surprised look on his face, MacDiarmid said: "Oh you mean capital N!" He then stood up, strode quickly to the blackboard and wrote the following: $(SN)_x$. not $(Sn)_x$. Actually, he first wrote $(Sn)_x$, then crossed it out and wrote $(S\underline{N})_x$, appropriately underlined for emphasis. The origin of the misunderstanding was immediately clear to me: Sn is the chemical symbol for tin, and everyone knows that tin is a metal — so Professor Heeger, what is the big deal!

After we got past this simple language problem, we quickly got down to a serious discussion. I pulled out a copy of the issue of *Physical Review Letters* that contained the publication by Labes and colleagues, and we quickly went over a few of the details. One might have thought that I should simply have opened the journal at the beginning of the conversation (and maybe I should have done!), but I had concluded that such a frontal approach would not be successful. The chasm between physics and chemistry was too great; without an understanding of the context of my interest in $(SN)_x$ as a quasi-one-dimensional synthetic metal, the discussion would have been doomed from the beginning.

MacDiarmid quickly realized that because of his early research in New Zealand, he was perhaps uniquely prepared to pursue the synthesis of this exotic material. A series of plans were made followed by joint meetings during the next few days with members of our two research groups. The Heeger–MacDiarmid collaboration (always in alphabetical order, even in the announcement of the Nobel Prize 25 years later) had been initiated.

MacDiarmid and I have the same first name, Alan; even the spelling is the same. When we are together, we always address each other as "Alan." Since, however, a reference to the "Alan–Alan collaboration" is meaningful to only a few close friends and colleagues, I will refer to Alan MacDiarmid by his last name.

The $(SN)_x/(Sn)_x$ confusion actually had positive consequences. We realized that we were embarking on a novel and exciting, but somewhat risky, interdisciplinary adventure. The $(SN)_x/(Sn)_x$ confusion demonstrated how little I knew about the chemistry and how little MacDiarmid knew about the physics of the new field that we were about to create. In an attempt to teach one another, to learn the languages of the two disciplines and to foster our nascent interdisciplinary collaboration, we began a series of regular meetings that were held in my office on Saturday mornings starting at 10 am. Sometimes, we discussed specific research results. Sometimes we worked together to complete a manuscript that reported some aspect of our early studies of the physics and chemistry of $(SN)_x$, including for example, the analysis of the crystal structure of the polymer, and the analysis of the golden reflectance and how the latter can be used quantitatively to obtain fundamental information on the metal physics of $(SN)_x$. Sometimes we simply discussed what we viewed as important opportunities for new research directions.

On one of those Saturday mornings, I decided that I would teach MacDiarmid about the basic physics of the metal-insulator (M-I) transition. The physics of M-I transition continued to be a focus of interest to condensed matter physicists throughout the 1970s; I was convinced that we had an important opportunity to have impact on the general aspects of this problem through the synthesis of new materials with the quasi-one-dimensional structures that are characteristic of linear polymers.

The physics of the M-I transition is not simple. In fact, theoretical studies of the M-I transition were at the forefront of the field. In 1977, Sir Nevill Mott and Philip Anderson were awarded the Nobel Prize in Physics; each made fundamentally important, but different, contributions to our understanding of the origin of the transition from metal to insulator and vice versa. Mott is known for his insight into the role of the repulsive interactions between like-charged electrons as a mechanism for localization of the electrons; when the repulsive interactions between like-charged electrons is large compared to the energy gained from metallic delocalization, the material becomes a magnetic insulator rather than a metal — the so-called Mott transition. Anderson was the first to show how disorder and the resultant multiple scattering of electrons from structural imperfections provide a mechanism for localization of electrons. When the disorder is large compared to the energy gained from metallic delocalization, there are no connected paths for electron transport, and the material becomes an insulator. Together, their work provided the foundation for understanding electron transport in disordered and strongly correlated "metals."

In order to simplify the physics in our Saturday morning meeting, I started by suggesting to MacDiarmid that we consider, as a model system, a linear chain of hydrogen atoms; I knew that this simple model contained the essence of the physics underlying the Mott transition. By now, MacDiarmid and I had developed a solid and open relationship. When we disagreed with one another or when one of us did not understand something, we did not hesitate to say so. To my suggestion that "we consider, as a model system, a linear chain of hydrogen atoms," MacDiarmid simply said "No!" When, somewhat taken aback, I asked him "Why?" he responded that "a chain of hydrogen atoms does not exist in Nature," period, end of story. If I were going to make progress with him, I had to come up with a model system that did (or at least could) exist.

I had suggested the linear chain of hydrogen atoms because of its simplicity. Since each hydrogen atom has only one electron, the "bookkeeping" is relatively simple, but the important role of the repulsion between electrons can be brought out and made clear. I needed the analogue of a chain of hydrogen atoms — a chain of repeated units each with one electron outside a filled core. Lithium atoms and sodium atoms have one electron outside a filled core, but again a chain of lithium atoms or a chain of sodium atoms does not exist in Nature.

Carbon has a valence of four; a carbon atom has four electrons outside of a filled core. These four electrons are responsible for the chemical bonds that lead to all of organic chemistry and most of bio-chemistry. A simplest example of a carbon-based linear chain is the polymer known as polyethylene. The repeat unit is the ethylene molecule, C_2H_4 or $H_2C = CH_2$, where the carbon atom is denoted by C, the hydrogen atom is denoted by H and the " = " denotes a double bond, each chemical bond comprising two electrons shared by the two neighboring atoms. In the process of polymerization to polyethylene, one of the two bonds between the two carbon atoms opens, and the resulting unpaired electrons react with those from two other ethylene units to form the polyethylene structure sketched in Fig. 1 below:

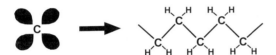

Eachof the four electrons on each carbon atom islocalized in covalent chemical bonds, each chemical bond comprising two electrons shared from the two neighboring atoms. There are no unpaired electrons in the repeat unit.

Fig. 1   The molecular structure of polyethylene.

In polyethylene, each carbon atom is bonded to two neighboring carbon atoms and two hydrogen atoms; the solid lines connecting the C and H atoms denote the chemical bonds. As a result, all four of the valence electrons (shown schematically on the left of the figure) are localized in strong covalent bonds like the carbon-carbon bonds in diamond. Polyethylene is, therefore, an insulator.

Polyacetylene, however, is fundamentally different than polyethylene — as different as three is different from four. In polyacetylene, the repeat unit is the acetylene molecule, C_2H_2, or HC≡CH, where the "≡" denotes a triple bond, each chemical bond again (and always) comprising two electrons shared by the two neighboring atoms. In the process of polymerization to the polymer polyacetylene, one of the three bonds between the two carbon atoms opens, and the resulting unpaired electrons react with those from two other acetylene units to form the polyacetylene structure. As shown in the Fig. 2 (next page), the

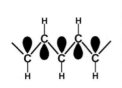

Three of the four electrons ion each carbon atom are localized in covalent chemical bonds, each chemical bond comprising two electrons shared from the neighboring atoms. The "teardrops" are non-bonded electrons in polyacetylene. These non-bonded electrons would delocalize and make polyacetylene a metal.

Fig. 2   Idealized molecular structure of polyacetylene.

idealized structure of polyacetylene, or $(CH)_x$, consists of a linear chain of C-H units, each with one electron (sketched in black) outside a filled core. Polyacetylene does, or at least could, exist.

Polyacetylene was sufficiently real to MacDiarmid that we went on to heuristically explore the C-H linear chain in the context of the Mott transition during one of our Saturday morning meetings. In fact, polyacetylene did exist. This polymer was first synthesized by the famous Italian chemist Giulio Natta in 1958, although I was not aware of Natta's work when I first discussed polyacetylene as a model system with MacDiarmid in 1975.

In this figure, the solid line connecting two carbon atoms (C-atoms) or one carbon and one hydrogen atom (H-atom) again denote chemical bonds. The solid tear-drop shape on each C atom indicates the fourth electron; this fourth electron is not involved in a covalent bond.

Sometimes students will naively ask me what is required to earn the Nobel Prize. I answer that one needs only to understand the difference between three and four — and in a very real sense that statement is true. Conjugated polymers, of which polyacetylene is the simplest example, are metals or semiconductors because of the concepts illustrated in the sketches of the molecular structures of polyethylene (all four valence electrons involved in covalent bonds) and polyacetylene (only three of the four valence electrons involved in covalent bonds). In the structure of polyacetylene shown in the figure above, all C-atoms are equivalent. Quantum mechanics tells us that these fourth electrons can delocalize along the poly-acetylene chain to form a metal. On average, there will be one electron on each C-atom, but if the C-atoms are close enough (and they are), these so-called pi-electrons will be free to move along the CH- chain; the molecular structure in the figure above could have the properties of a metal.

Shortly after our Saturday morning sessions on the metal-to-insulator transition as viewed through the conceptual example of polyacetylene, MacDiarmid went to Japan on a lecture tour. He visited many Japanese universities and typically gave one or more lectures. He departed from Philadelphia for Japan loaded with new and important data on the chemistry, the structure, the spectroscopy and the metallic properties of $(SN)_x$. He carried with him sealed glass vials containing small golden crystals of $(SN)_x$ and other vials containing golden thin films of $(SN)_x$ deposited onto glass slides. He also had photographs of these beautiful golden crystals and golden films reproduced on color slides for projection during his lectures.

MacDiarmid was a very "visual" person; color was important to him. He was very proud of his golden crystals and his golden thin films of $(SN)_x$. Chemists are generally sensitive to color; they often make specific mention in their publications of the color of a compound and of the changes in color that occur during a chemical reaction. I am not usually so impressed with color; I get my kicks from the graph of the reflectance or absorbance of a compound as a function of the wavelength of light or as a function of the photon energy. The reflectance and absorbance determine the color, but there is much more quantitative information contained in the graphs of reflectance and absorbance than that contained in a description of the color.

After his lecture on poly(sulfur-nitride) at the Tokyo Institute of Technology, MacDiarmid was invited to tea by his host and a small group of chemistry colleagues. During the ensuing discussion, he was told that a young chemist at the Tokyo Institute of Technology, Hideki Shirakawa, had some similarly impressive films of a polymer, but that Shirakawa's films were "silvery" rather than golden. Because MacDiarmid expressed keen interest in seeing some of these silvery films, Dr. Shirakawa was contacted and invited to join the group. In fact, Dr. Shirakawa had not attended MacDiarmid's lecture. Poly(sulfur-nitride) is an inorganic material whereas Shirakawa's interests were focused on organic (carbon-based) polymers; such is the nature of scientific specialization. After being introduced to MacDiarmid, Shirakawa showed Alan a number of glass vials containing free standing silvery films. Shirakawa's vials contained films of polyacetylene.

I learned about Shirakawa's silvery films of polyacetylene the day following MacDiarmid's return to the University of Pennsylvania. He was excited with the silvery films and with what he had learned during his brief discussion with Shirakawa. We were intellectually prepared; we had discussed the electronic structure of polyacetylene as an idealized linear polymer during our Saturday morning meetings. Within a few minutes, we decided that we should bring Dr. Shirakawa to the University of Pennsylvania to work with us, and we should do so as soon as possible.

We came to this decision without a great deal of discussion even though we both knew that we were embarking on a very risky project. Our work on poly(sulfur-nitride) had been solid as a rock; we had expertise in the synthesis (because of MacDiarmid's early experiences as a student in New Zealand), we had successfully grown well-ordered crystals of $(SN)_x$, and we had X-ray diffraction data that told us precisely where the S and N atoms were located in the linear chain structure. Not one of the items on this list was true for poly-acety- lene. MacDiarmid was an inorganic chemist, the silvery films were only partially crystalline, and there was no definitive information available on the molecular structure or the arrangement of the supposed linear chains within the material. Despite this lack of expertise and the incomplete state of our knowledge, we did not hesitate at the water's edge; we decided to jump in.

After the award of the Nobel Prize in 2000, several people made comments to me that we had carried Shirakawa across the goal line, implying that he did not deserve this singular honor. I disagreed then, and I disagree now. Shirakawa discovered the synthesis of the silvery films of $(CH)_x$ that opened the entire field of semiconducting and metallic polymers. His careful work on the chemical analysis and on the molecular structure through analysis of the infrared signatures, his initial X-ray diffraction data, and his painstaking electron microscopy studies of the morphology built a foundation upon which all of our early work rested. It was the strength and solidity of that foundation that enabled us to go forward. Without that foundation, the scientific communities would not have condoned the rapid progress that ensued. After Hideki came to join us at Penn, he was a key player in the discovery of the effect of charge transfer doping on the electrical conductivity of polyacetylene. Without any doubt, he deserved his share of the Nobel Prize in Chemistry.

The story of how Shirakawa discovered the synthesis of the silvery films of $(CH)_x$ that opened the field has been often retold. The simplicity of the linear chain structure of polyacetylene had not gone unnoticed. Although first synthesized by Natta in 1958, the material that Natta obtained was an uninteresting black powder. Natta had more important work to do than waste his time on this intractable black powder, so the polymer of acetylene was bypassed for more than a decade. Shirakawa began his work on the polymerization of acetylene in the early 1970s. The transition from the uninteresting black powder to the silvery films that attracted our attention was the result of an accident in the laboratory — more precisely, the result of a misunderstanding in the laboratory. Acetylene is polymerized through the use of a catalyst from the group known as Ziegler–Natta catalysts, after the two Nobel Laureates who discovered them. In Natta's early work as well as in Shirakawa's early work, the catalyst was used in the very dilute, millimolar concentrations typical of catalysts. Shirakawa hosted a Korean visitor in his laboratory for a short period of time. He carefully instructed the visitor on the method of synthesis, but because of the Korean's limited knowledge of Japanese, the visitor misunderstood the instructions; the catalyst was added to the reaction mixture at a concentration that was 1,000 times higher than used in Shirakawa's earlier work. Instead of the expected black powder, the visitor came to Shirakawa with a mess; the solution had formed a gel that he found impossible to work up or to purify. I am certain that 99 out of 100 chemists would have told the visitor that he made an error, to throw the mess away, and to re-do the synthesis "correctly" according to the instructions given to him. Hideki Shirakawa was the one out of 100 who had the foresight and the curiosity to pursue the matter. He did, and because of his efforts, he created the key that opened the door to the discovery of conducting polymers.

We had decided to bring Shirakawa to Philadelphia, and he had indicated that he would like to join us. We needed money to make this happen. Dr. Kenneth Wynne, for many years the Director of the Polymer Program at the Office of Naval Research, came to our aid. He was willing to back us on this endeavor despite the fact that we had no track record of any kind in the area of organic polymers and despite the fact that conducting polymers (our stated goal) did not exist. Risk is involved at every level at the beginning of anything that is truly new. In a way finding the resources to back scientific investigation is like finding an angel to back a theatrical production. Both are calculated risks but still risks. Dr. Wynne was indeed our funding angel.

In 2000, Ken Wynne joined us in Stockholm and witnessed the Nobel Prize ceremony; he deserved to be there.

The actual structure of polyacetylene is not that shown in Fig. 2. If it were, polyacetylene would be a metal. Polyacetylene is not a metal, but a semiconductor with a relatively small energy gap. The bond-alternated structure sketched in Fig. 3 has a lower energy than the structure

The actual molecular structure of polyacetylene consists of series of short (or double) bonds and long (or single) bonds. The neighboring unpaired electrons of Fig. 2 pair to form relatively weak so-called π- bonds.

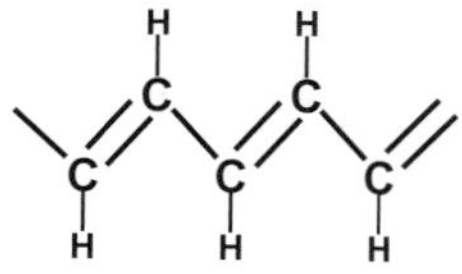

Fig. 3   The bond-alternating molecular structure of polyacetylene.

sketched in Fig. 2 (and is therefore the structure of real samples) because of the Peierls instability of one-dimensional metals.

The bond alternating structure sketched in Figure 3 was predicted in general terms by the theory created by Professor Peierls. The bond-alternating structure is, however, not a theoretical conjecture. It is a fact proven by a combination of X-ray diffraction and nuclear magnetic resonance experiments carried out in the 1980s.

Shirakawa arrived in Philadelphia accompanied by his wife, Chiyoko, and his two sons in late October, 1976. After the few days required to settle into an apartment and a new (although temporary) life, Hideki Shirakawa, Alan MacDiarmid and I sat together in my office in the Laboratory for Research in the Structure of Matter to formulate a plan for our research on polyacetylene. We had the key idea during a relatively short period of discussion one afternoon in early November 1976. We postulated that we could convert semiconducting polyacetylene into metallic polyacetylene by charge transfer doping — that is, by inserting between the weakly interacting polymer chains molecules that function as donors of electrons or molecules that function as acceptors of electrons. Our subsequent discovery of metallic levels of electrical conductivity in polyacetylene, published in 1977, demonstrated that our ideas were both true and revolutionary. The risk that we had taken resulted in the creation of a new interdisciplinary field at the boundary between the traditional disciplines of chemistry and physics: the field of semiconducting and metallic polymers.

Had we really created this new field? Were we really first? The answer to such questions was clearly and eloquently first stated by Sir Isaac Newton: "If I have seen further, it is by standing on the shoulders of gaints." Many chemists and physicists had touched upon the subject of conjugated polymers. There were reports in the literature of modest electrical conductivities. We were aware of some of these and ignorant of others. I will not even attempt to list those earlier "giants" because I would surely leave out important names that should have been included. Our interdisciplinary approach, the strength, breadth and solidity of our results, the new theoretical concepts that we introduced, and the firm scientific foundation that we built all contributed to what turned out to be the creation of the field of semiconducting and metallic polymers.

When MacDiarmid, Shirakawa and I were awarded the Nobel Prize in Chemistry nearly 25 years after the initial discoveries, the citation read "for the discovery and development of conducting polymers." The actual discovery occurred over a few short weeks; the subsequent development of the science, the materials, and the applications required nearly 25 years. Twenty-five years from discovery to the award of the Nobel Prize is a long time, but not unusually long for Nobel Prizes. In our case, time was required to make certain that the science was right and that the impact was real. Time was required for other scientists to verify the initial discovery and to broaden it to a larger class of materials beyond polyacetylene. Time was required to develop the new theoretical concepts that were required to describe the fundamentally new phenomena associated with the field of conducting polymers — and time was required for the scientific community to accept these new theoretical concepts. Time was required to achieve stable materials, and time was required to achieve materials that could be processed into useful forms. Time was required to invent and develop the unique applications for which conducting polymers offered special advantages. The development of science and technology takes time.

The twenty-five years between the initial discovery and the Nobel Prize was certainly not a boring period. These years were filled with the discovery of new examples of conjugated polymers, with the discovery of new and unexpected phenomena, and with the creation of a knowledge base for the field of conducting polymers.

As demonstrated in our early publications on the electrical conductivity of polyacetylene, we had developed effective methods for introducing donor or acceptor molecules into the polymer. However, we had problems controlling the reaction. Typically, within a few minutes the reaction went to completion at the highest concentration of dopants. Although this gave us access to experiments that could probe the properties of the metallic state, I wanted to be able to control the reaction and to be able to stabilize the electronic properties at any desired intermediate level of doping.

I was bemoaning my frustration on this problem to MacDiarmid at lunch one day. We were at a McDonald's restaurant within the Children's Hospital in the Medical Complex adjacent to the University of Pennsylvania campus. While munching our lunch and discussing between bites, we had the idea. It was indeed like turning on a wall switch — all at once there was light. We realized that we could add or withdraw electrons using the methods of electrochemistry. It was a moment of exciting discovery. We hurried back to the lab, found Paul Nigrey (a graduate student who had been working on the synthesis of polyacetylene), and described the concept to him. Within a few hours, he demonstrated that the idea was correct. Electrochemical doping quickly became our method of choice. The electrochemical methodology not only provided samples for detailed study, it also established yet another sub-field within chemistry: conjugated polymers as electrochemically active materials.

Subsequent measurements on stretch aligned samples of polyacetylene yielded electrical conductivity values approaching that of copper and a tensile strength greater than that of steel — a remarkable accomplishment, and one that certainly provides stimulus for the technology of plastic electronics. Thus far, such high conductivity values have been demonstrated only in doped polyacetylene. However, I am confident that when polymers such as polyanilene are macroscopically aligned with a similar level of structural integrity, they, too, will exhibit truly metallic levels of electrical conductivity.

New phenomena are always interesting; the discovery of new phenomena is at the heart of experimental science. New theoretical concepts are equally important, especially when they lead to the prediction of specific experiments that are capable of verifying (or falsifying) the new concepts. In this context, the theoretical development and subsequent experiments on "solitons" in polyacetylene were particularly important to the development of the field of conducting polymers.

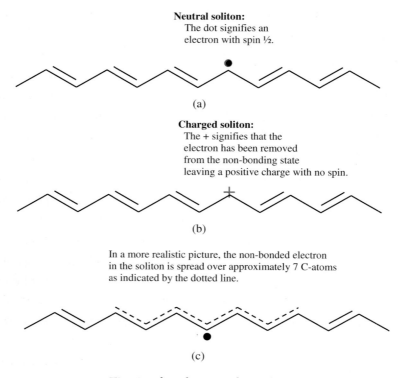

Neutral soliton:
The dot signifies an
electron with spin ½.

(a)

Charged soliton:
The + signifies that the
electron has been removed
from the non-bonding state
leaving a positive charge with no spin.

(b)

In a more realistic picture, the non-bonded electron
in the soliton is spread over approximately 7 C-atoms
as indicated by the dotted line.

(c)

Fig. 4   The soliton in polyacetylene.

As shown in the Fig. 4, there are two "forms" of polyacetylene,
the one on the left with the double bond pointing down and the one on
the right with the double bond pointing up. Because the two kinds are
identical by symmetry, they must have identical properties. Nevertheless, the two structures are distinct and separately stable. When one
connects the two different forms, however, something quite remarkable happens. Simple counting again gives the essential features of the
answer: a single non-bonded electron, shown schematically by the dot,
appears at the boundary between the two forms. On the C-atoms on the
left and on the right, the four electrons on each C-atom are involved
in chemical bonds (including the C-H bond not specifically shown in
this shorthand notation). On the C-atom at the boundary, however, only
three electrons are involved in such chemical bonds; the fourth electron resides on the boundary C-atom without being bonded to either

of the neighboring C-atoms; the fourth electron is in a non-bonded configuration. Because single bonds are longer than double bonds, the boundary — the soliton — consists of a structural distortion that localizes an electron in a non-bonding state. This boundary can be anywhere — since the structure on the left is energetically equivalent to the structure on the right, the boundary can be moved without any change in the total energy. Thus, the soliton in polyacetylene can be anywhere on the polymer chain, and it is therefore inherently mobile.

The discovery of solitons in polyacetylene was a major event; it provided a fundamental theoretical structure for the field of semiconducting polymers. In the Spring of 1978, I gave a lunchtime seminar to my colleagues in the Solid State Group at Penn. During this informal seminar, I discussed interesting and puzzling new data that we had recently obtained from polyacetylene: a spin resonance when there should not have been one and intense infrared modes that appeared upon doping the polymer. Toward the end of my presentation, I gave a somewhat "fuzzy" picture of a boundary-like structure similar to that shown in Fig. 4.

Bob Schrieffer made a few comments to me after the seminar. He seemed to be on the verge of understanding something, but I could not get it out of him. The next day he came to me with a complete picture of the soliton in polyacetylene and the implications of this new "particle" on the electronic structure of polyacetlyene. Schrieffer and his student, Wu Pei Su, then proceeded to work out the theory of solitons in polyacetylene in detail. The subsequent Su–Schrieffer–Heeger (SSH) publications provided a theoretical framework for the field of conducting polymers.

As shown in Fig. 4, the neutral soliton has charge zero and spin ½ whereas the positively charged soliton (b) has charge +e but no spin ("e" is the electron charge, $e = 1.6 \times 10^{-19}$ Coulombs, a fundamental constant of physics). There is also a negatively charged soliton with two electrons in the non-bonded soliton state.

Do not be frightened by these very simple diagrams. A straight line between two atoms signifies a chemical bond. The important point of the diagrams in this figure is that by reversing the bond alternation, one creates an electron in a non-bonded state, i.e. not bonded to either of its neighbors. As a result, this soliton structure with the associated non-bonded electron can move and transport charge along the polymer chain. This was a new "kind" of charge transport and therefore particularly interesting.

Electrons have spin (s = ½) and charge e = 1.6×10^{-19} Coulombs, but the soliton exhibits a *reversal* of the famous spin charge relationships identified by the Physics Nobel Laureate Wolfgang Pauli during the early development of quantum mechanics. Our predicted and experimentally demonstrated reversal of Pauli's fundamental spin-charge relationships caused quite a stir within the physics community.

Solitons in Polyacetylene

	Charge	Spin
Neutral soliton	0	± ½
Positively charged soliton	+ e	0
Negatively charged soliton	− e	0

The predictions of the SSH theory were subsequently confirmed in detail through a series of experiments carried out in laboratories all over the world. The confirmation of these theoretical predictions did not come easily; it is always difficult in science to "get it right." The soliton in polyacetylene and the generalization of this concept to other semiconducting polymers were topics of contention, argument and controversy for several years. The initial SSH theory was purposely simplified to bring out the fundamentally new features associated with the schematic structure shown in Fig. 4. The interactions between electrons were ignored in the initial SSH model, but were subsequently included as the field developed. The most important predictions of the SSH theory are robust; they originate from a combination of fundamental symmetry and counting, as outlined above. The reversed spin-charge relations summarized above were verified as were the electronic structures and energy levels of the neutral and charged solitons.

As I have emphasized, the greatest pleasure of being a scientist is to have an abstract idea, then to do a series of experiments that demonstrate that the idea was correct; that is, that Nature actually behaves as conceived in the mind of the scientist. This process is the essence of creativity in science. The story of "solitons in polyacetylene" is without doubt one of the best examples of that experience.

When we started our research on polyacetylene, the basic concepts that define semiconducting polymers and metallic polymers were not understood. Despite the solid foundation built by Shirakawa, the existing materials were poorly defined and incompletely characterized. According to the standards of physicists, they were "dirty materials." Moreover, the ultra-high magnification photographs published by Shirakawa (obtained by transmission electron microscopy) showed a complex, porous, fibrillar structure that I found scary to look at, let alone to contemplate as the basis for deep fundamental insight or as a potentially useful material. How could electrons move efficiently through that mess? To make matters worse, polyacetylene was unstable; its properties degraded rapidly when exposed to air. Some of our early samples even spontaneously ignited when exposed to air. As a result of this instability, all of the research required handling the samples in a controlled atmosphere glove box or storing the samples in vacuum in sealed vials. My physicist friends thought I was crazy.

Why would I take such a serious risk? There were three reasons:

(1) Conducting polymers did not (could not?) exist.
(2) Conducting polymers promised a unique combination of properties not available from any other known materials.
(3) I love the feeling of being out in front and leading the way.

The first expresses an intellectual challenge, the second expresses a promise for utility in a wide variety of applications, and the third is just part of what I am.

We tried to mitigate the risk. In science, the way one often mitigates risk is to utilize simplicity. We began with polyacetylene, $(CH)_x$. Polyacetylene has a remarkably simple molecular structure. We published our first paper outlining the initial results in *Chemical Communications* in early 1977. The first publication that focused on the electronic properties of polyacetylene was published in *Physical Review Letters* later that same year. There was no risk in the title: "Electrical conductivity in polyacetylene." We had discovered that the electrical conductivity of polyacetylene could be increased and controlled from that of an insulator to values approaching those of metals. In these early experiments, we demonstrated an increase in electrical conductivity by a factor of approximately one hundred million. Nevertheless, I managed to get into the abstract and into the first paragraph the claim that we had discovered an example of the "insulator to metal transition" in this system. That claim was not obvious, it was an interpretation of the data. We thought we had it right and, as it turned out, we had gotten it right, but there was risk in presenting that interpretation.

The Heeger–MacDiarmid–Shirakawa collaboration was ideal in every way. We simply worked well together. We trusted each other and never had a serious disagreement. Fortunately, everything went well from the very beginning and went well with remarkably rapid progress. This was a "perfect" collaboration. We did not keep this perfect collaboration together beyond a few years. Shirakawa returned to his position in Japan after one year. I moved to Santa Barbara approximately six years later. We always remained friends and colleagues, but the intense collaboration of the early years ended as we separated to different parts of the world.

Within a relatively short period of time following the initial discovery of conducting polymers, it became clear that we had created a new field and provided a fundamental theoretical foundation for that new field:

(1) We had indeed demonstrated the conducting polymers could and did exist!
(2) The demonstration that conducting polymers offered a unique combination of properties not available from any other known materials was yet to come.
(3) I was certainly out in front and leading the way.

The years between 1976 and 1982 were among the best years of my life as a scientist with many "perfect days" along the way.

Chapter 12

The Nobel Prize in Chemistry

The telephone rang at 5:45 am. I knew — suspected, hoped that I knew — who was calling. The date was October 10, 2000, the week during which the Nobel Prizes for the year 2000 were being announced. The Prize in Medicine and Physiology had been announced in Stockholm the day before. Today was to be the day for the announcements of the Nobel Prize in Physics and the Nobel Prize in Chemistry.

I was aware because my colleague scientists had been telling me for years that I would one day receive the Nobel Prize — the most prestigious prize in science — for my work in conducting polymers and the physics of quasi-one-dimensional conductors. There had been rumors over the years. The Nobel Foundation had sponsored a symposium on the topic of our work in 1991; such a Nobel Symposium is sometimes, but not always, a precursor to the awarding of the Prize.

In fact, year after year nothing had happened.

The rumor had been particularly strong in 1998. Ruth and I were in Hokkaido (Japan) during the "announcement week"; I was participating in a scientific conference there. Our hosts alerted the local press that Professor Alan Heeger was in town and that "tomorrow" I would be notified that I had received the Nobel Prize. It did not happen.

My daughter-in-law often slept with the radio on and an earphone in her ear. She had a moment of uncertainty early that morning in 1998 when she heard that a professor from the University of California at Santa Barbara had been awarded the Nobel Prize in Chemistry. But, alas, the professor was not Alan Heeger but my colleague Walter Kohn, a physicist and a long-time friend of mine. Walter's well-deserved Prize

was the first for UC Santa Barbara and therefore a cause for great celebration in Santa Barbara. We missed all the excitement because we were
far away in Hokkaido.

After the false rumor in 1998, I think I gave up; it seemed that
the attempts to convince the Nobel Committee and the Swedish Academy of Science had failed. There is typically a "window in time" for any
such recognition. There are many truly outstanding scientists who have
made ground-breaking discoveries, any of whom might well be properly
recognized with the Nobel Prize. Despite the optimism of some of my
scientific colleagues, the probability that I would actually receive the
telephone call was small.

I was, therefore, indeed surprised — but quickly aware — when
the telephone rang at 5:45 am on October 10, 2000. We were in bed
and had been sound asleep. The telephone in our bedroom is on Ruth's
side of the bed. She answered after the awakening ring and said to me
in a sleepy voice that "there is someone on the phone from Sweden."
I responded: "Don't hang up!"

On the other end of the call was the President of the Swedish
Academy of Sciences. I do not actually remember the words that he
said, but I got the essential points. I had been awarded the Nobel Prize
in Chemistry with my two colleagues, Professor Alan G. MacDiarmid
and Professor Hideki Shirakawa "for the discovery and development of
conducting polymers"; that is, for our discoveries of polymers that have
the electronic and optical properties of semiconductors or even those of
metals. At the end of the call, the telephone in Sweden was passed to a
friend whose voice I recognized. He, too, congratulated me. I realized
after the end of the call that his familiar voice was the "reality check";
this was not a hoax.

Why 5:45 am? The answer is simple, the announcement of the
Chemistry Prize is made during a press conference in Stockholm at
3:15 pm; that is, 6:15 am in the early morning in Santa Barbara. The call
came at 5:45 am to give sufficient time for us to contact our family and
special friends. We were told, however, that after 6:15 am, we would

not be able to make outgoing calls because of continuous incoming calls from the press. In the 20 minutes remaining, we were able to contact our sons and the closest members of our family.

The fact that I had been awarded the Nobel Prize in Chemistry rather than in Physics was the first and most important piece of information. As said by Lord Rutherford on the similar occasion in 1908 and for the same reason (he, too, was a physicist who was awarded the Nobel Prize in Chemistry): "What a remarkable transformation!" I had been born to be a physicist, I was trained as a physicist, I think like a physicist, and my sons always joke with me that I even look like a physicist. But life is indeed interesting: We are what we have become, and on October 10, 2000, I learned that I had become a chemist.

The Nobel Week in Stockholm was magical. We were treated as if we were either royalty or rock stars or both. Each Laureate and spouse had an assigned limousine and driver and a person from the Swedish Foreign Service to attend to our needs and desires. We were met at

Ruth and Alan on the morning of October 10, 2000 shortly after receiving the telephone call from the Nobel Foundation.

the airport by the President of the Nobel Foundation and taken to the Grand Hotel (it is truly Grand!), where we settled into a room overlooking the harbor. Of course nothing is perfect. The closet was too small for Ruth's gown and all her clothes, but the management without hesitation gave us a double room and brought in a large rack on which to hang her clothes.

The highlight for me and for Ruth was of course the award ceremony. She sat close to the front of the hall with her four-year-old granddaughter, Alice, at her side. Alice was restless, but was amused by my making faces at her during the somewhat long ceremony. She got visibly upset only when the speeches were given in Swedish. This was her first experience with a foreign language, and she could not understand what was going on.

The banquet following the ceremony is held on the lower floor of the City Hall. It was beautifully decorated, and the ceiling was lighted with changing colors and patterns. There were two Laureates from UC Santa Barbara that year. Herbert Kroemer received the Physics Prize and I received the Chemistry Prize. Herb is a few years older than I am,

Alan receiving the Nobel Prize from Carl XVI Gustaf, King of Sweden.

Ruth and Alan in a quiet moment of celebration after the Nobel Ceremony.
Ruth looked more beautiful than the Queen. Alan is enjoying some single
malt Scotch whisky.

and although we are not related we are both old men with white hair and
beards. We were mistaken for one another again and again. His picture
appeared in the Stockholm paper with my name under it (or perhaps it
was me with his name under it, I am not certain).

Ruth was truly in the "spotlight" at the banquet and at the more
private dinner with the King and the Royal Family the following night.
The Laureates and spouses came into the Banquet Hall, two by two,
down a very long flight of steps accompanied by triumphal music. At
the top of the stairway, we were arranged so that each would arrive at
his or her assigned seat at the very long center table with the King and
Queen in the center, but on opposite sides. Ruth sat next to the King, but
came down the long stairway on the arm of my look-alike colleague Herb
Kroemer — confusing for the photographers and the other hundreds of
guests at the banquet. Care was taken to assure that all the details were
correct. Ruth is allergic to bivalve shellfish (mussels, clams and oysters,
etc.). There were bivalve shellfish in the first course, but when I looked
over at her, I could see that her plate was different; the shellfish had
been replaced by a salad.

Alan shows the medal to his granddaughters Alice and Julia after the ceremony.

The dinner the following night was held in the Royal Palace and was a more intimate affair with only the Laureates and spouses, the Royal Family and a few obviously important, elite members of Swedish society. We gathered in an entry room and chatted with one another and our new Laureate friends for a few minutes. Then suddenly a very tall man in medieval costume and wearing an even taller hat and carrying a huge mace that he pounded on the floor at every step was standing in front of Ruth. Somehow he indicated to her that she should follow him. He took her to the King, who was to escort her to dinner. She was poised and beautiful in a long blue dress. She bowed to him, and the King took her arm. She then took one step and slipped on the polished marble floor and found herself in the arms of the King. Embarrassed, but laughing, she retained her composure and proceeded to dinner. The main course was venison. As her dinner companion, the King regaled her with stories of how he had personally shot the deer that provided the venison.

Some months later, after we had returned to Santa Barbara, I was awarded the "Pioneer of the Year" award by the local Chamber of Commerce. It was quite an elaborate affair with approximately 200 people in attendance. I received the award from Michael Towbes and quickly commented that although I had received another award recently (laughter!), I was especially honored to receive this award from Michael, who is well known in Santa Barbara as a businessman and a wealthy and very generous philanthropist. When thanking Michael, I stated to the audience: "I think that Michael is smarter than the King, he is certainly better looking than the King, and I think that he probably has more money than the King!" This brought the house down and greatly pleased Michael Towbes.

The Nobel Medal is gold, and about 3 inches in diameter. On the front is the image of Alfred Nobel, with his birth year (MDCCCXXXIII = 1833) and the year he died (MDCCCXCVI = 1896) inscribed beside him in Roman numerals. On the back are two beautiful female figures and my name, together with the year of the award of my Prize at the bottom (MM = 2000). I was pleased with the simplicity of the Roman numerals that designated the date of my Prize. Because of the complexity of the Roman numerals, it would be 1,000 years before such a simple inscription would be seen again (MMM = 3000)!

The Nobel Prize is generally considered by scientists to be the premier award for scientific accomplishment. That is why the award of the Prizes gets such significant press coverage each and every year. The Prize is the responsibility of the Swedish Academy of Science. Each year, committees are appointed to supervise the recommendations to the Academy of the scientists who were nominated that year for the award of the Prize. The final decision is made by the Swedish Academy of Science.

I now know several Swedish colleagues who have served on either the Physics Nobel Committee or the Chemistry Nobel Committee in years past. They take great care to see that they are making the correct decision. They request and receive nominations from prominent scientists from all over the world. In addition, every professor from a university in one of the

Scandinavian countries is allowed to nominate a candidate every year if his appointment is in one of the five specific disciplines designated by Alfred Nobel. When the Committee (for example, the Chemistry Committee) has narrowed the list to a final few, they ask for independent reports and analyses from external experts who have earned international reputations for their contributions to related science. They even make inquiries as to exactly who was in the laboratory at the time of the discovery. The evaluation and decision process are taken very seriously. Although there might have been mistakes made over the 115 years that the Prize has been given, such mistakes are indeed very rare. I was told by one of my Swedish colleagues that they consider their job as equivalent to writing the history of the evolution of science. They take their responsibilities very seriously indeed, and they rarely make mistakes.

Physics, chemistry and biology all originated in man's fascination with natural philosophy. There are many problems in science that are interesting, but success in terms of impact fundamentally relies on doing something that is truly new and stimulates the growth of a new field within the larger scope of science. This is true in any area of human endeavor. "Breaking new ground" that is sufficiently exciting to others that they follow your lead is the key to high-level success in science.

Science is basically difficult not because the concepts are difficult (although they sometimes are) but because the scientist must "get it right." If I go skiing on a very steep slope and break my leg, my broken leg will eventually mend. But one cannot make mistakes in science; the scientist must get it right. The need to "get it right" implies that creating new scientific knowledge involves risk. Moreover, if one is seeking to find new discoveries, one must, by definition, venture out in new and previously unexplored directions. All such ventures into new directions involve risk. Of course, mistakes are made; I have made more than a few myself. When one realizes that such an error has occurred, the goal is to clean up "one's act," and, subsequently, "get it right."

Exploring new directions is the first step toward creativity and discovery. This certainly turned out to be true in my scientific life. I (the physicist) shared the Nobel Prize in Chemistry with Alan MacDiarmid, who was an inorganic chemist, and with Prof. Hideki Shirakawa, who was a polymer scientist; a truly interdisciplinary trio. I had no previous knowledge of the chemistry or physics of polymers. Neither did Mac-Diarmid. As an inorganic chemist, his research life had been focused on reactions and materials from inorganic chemistry, i.e. chemistry that is not based on carbon. Fortunately, Shirakawa was indeed a polymer chemist. Each of us learned a great deal from one another.

An essential feature of an interdisciplinary collaboration of this type is to understand one's role in the collaboration (that is, to respect the special expertise of each individual), to be sensitive to the day-to-day problems of one's colleagues and to be generous with your time and effort to help solve such problems even if it is not at that particular moment your personal highest priority. I was aware of these issues and was the leader of the collaboration. I always respected the expertise of my colleagues and made a point of helping when help was needed. My colleagues responded in kind.

Prior to our work there were three "generations" of polymer materials. The first generation consisted of natural polymers: spider webs, silk and even leather are examples. Although these polymers have been around and used for thousands of years, synthetic chemists had no understanding of what polymers really were; they were typically considered as the "gunk" in the bottom of the flask that resulted from some unknown side reaction. The concept of polymers as long chain macromolecules comprising a single molecular unit repeated many, many times was proposed by Hermann Staudinger in the 1920s. His subsequent experimental verification of the concept was a true break-through. Staudinger received the Nobel Prize in Chemistry for his discoveries in 1953.

The discovery of nylon and the ability to spin relatively strong, thin, synthetic fibers of nylon by Wallace Carothers in 1934 led to the second generation of polymers. The DuPont company, where Carothers made his discoveries (at the DuPont Experimental Station in Wilmington) quickly commercialized nylon fibers to make a new generation of sheer hose for women (an immediate and long-lasting commercial success!) and later as the material in parachutes during World War II. Carothers was elected to the National Academy of Sciences and would likely have received the Nobel Prize in Chemistry, but he died tragically (suicide) as a young man. I do not know enough about the mental history of Carothers to have insight into his suicide, but research has shown that bipolar disorder and schizophrenia correlate with high creativity and intelligence. The creative mind is a delicate beast.

The third generation, the "plastics" that are so important to our society today, was made possible by the discoveries of Karl Ziegler and Giulio Natta, who were awarded the Nobel Prize in Chemistry in 1963 "for their discoveries in the field of the chemistry and technology of high polymers." The catalysts discovered by Ziegler and Natta were responsible for the large-scale industrialization of polymers.

Paul Flory provided the basic theoretical understanding of polymer physics, and he paved the way for the development of thermo-plastics that are used as structural materials so important in our modern world; his classic book, *Principles of Polymer Chemistry*, is still a definitive source. Flory was awarded the Nobel Prize in Chemistry in 1974.

None of these first three generations of polymers, however, is interesting from the point of view of electronic materials. They are insulators and broadly important as structural materials, with products ranging from plastic bags and soft-drink bottles to bullet-proof vests. Many of the parts in today's automobiles are made of plastic. The fuselage of the new Boeing 787, United Airline's "Dreamliner" is made of carbon-fiber-reinforced plastic. Plastics are used in quantities larger than any other class of materials, larger in volume and even larger in weight.

The materials that my colleagues and I discovered, semiconducting and metallic polymers, brought electronic function into the area of polymer science. Semiconducting and metallic polymers are the "Fourth Generation of Polymeric Materials." They are electronically active, they offer both the electronic/optical properties of metals and semiconductors <u>and</u> the processing advantages and mechanical properties of polymers.

The different colors in the image correspond to different semiconducting and metallic (the darkest one) polymers, all in solution, with different molecular structures and, therefore, different optical absorption spectra and different colors.

The electric/optical properties of metals and semiconductors and the processing advantages and mechanical properties of polymers.

In the early days following our initial publications, the number of scientists who initiated research in the field of semiconducting and metallic polymers grew rapidly as people realized the importance of and the opportunities within this new field. They were attracted by the implications to fundamental science and the opportunities for innovative applications. One measure of the growth of the field and the impact of our discoveries is simply the number of times my publications have been cited by others. Each citation means that a group of scientists had made sufficient progress in their own research in the area of semiconducting and metallic polymers that they cited one of my publications when they published their new results. The record of the number of citations is readily available on the internet at websites such as Google Scholar (scholar.google.com). The number of citations of my published articles has grown steadily each year following our initial publications (in 1976) from approximately 100 in 1975 to 10,268 in 2013. The total number of citations of my publications during this period (1975–2013) is greater than 135,000. Thus, with our discovery of semiconducting and metallic polymers, we created a new field of science at the boundary between chemistry and condensed matter physics. Although this comparison is over simplified, chemistry focuses on the creation of new molecules and new materials, and condensed matter physics focuses on exploring the novel properties of such new materials with the goal of understanding these properties based upon fundamental principles. For example, why and how do metals conduct electricity?

Despite my track record of having received the Buckley Prize from the American Physical Society and the Nobel Prize in Chemistry, subsequent publications are not accepted by the community without a thorough vetting and checking in the laboratories of many scientists. One of the most important aspects of science is that it can be "falsified." A result is only accepted as true after it has been reproduced and carefully studied by others, whether the publication was authored by a prize-winning scientist or by a young beginning Assistant Professor. The referee reports are often critical, sometimes for good reason, sometimes because the referee did not understand the point under question. In what is perhaps my favorite referee report, the referee states that a specific paragraph in the manuscript is "spherical nonsense; i.e. it makes no sense from any point of view!" I read the paragraph again and agreed with the referee. Needless to say, I deleted the paragraph from the manuscript.

Referee reports are typically annoying to the authors because they usually require additional research to clear up some point of contention. In my experience, however, the referee system generally leads to improved manuscripts with minor errors corrected and with the scientific results presented with greater clarity.

Now, nearly four decades later, the discovery of metallic conductivity in polyacetylene is well known and often used as a highly successful example of the importance of interdisciplinary research. When we started this journey, however, the basic concepts that define semiconducting and metallic polymers were not understood. The existing materials were poorly defined and poorly characterized. Even after our initial discoveries were published, such an interdisciplinary endeavor was subject to the risk of being a "bastard child" that would not be accepted by either parent. In 1976, the creation of our truly interdisciplinary collaboration was bold and risky.

We were, of course, not the first scientists to take such risks. A classic and well-known example of risk and creativity is the great discovery by Albert Einstein of the quantum of light, the photon. In a six-month period from March through September in 1905, Einstein changed the world of physics. He discovered (invented) the photon, he published his Theory of Relativity, and he demonstrated the reality of atoms though his statistical theory of Brownian motion. In a separate paper, he deduced the famous equation, $E = mc^2$, as a consequence of his earlier paper on relativity.

In 1914 — nine years after Einstein's paper on the photon — Max Planck, the greatest physicist of his day — proposed the younger Einstein for membership to the Prussian Academy of Science. In his speech to the Academy, he praised Einstein's contributions to science, but he held back on his praise of Einstein's proposal of the photon. Planck said in this speech to the Academy: "That Einstein may sometimes have missed the target in his speculations, as for example, in his hypothesis of light quanta, cannot really be held too much against him, for it is not possible to introduce fundamentally new ideas, even in the exact sciences, without occasionally taking a risk."

Many people think that Einstein received his Nobel Prize for the Theory of Relativity. That is not true — the Nobel Prize in Physics was given to Einstein in 1922 for the discovery of the photon. The concepts of Einstein's initial publication on relativity (the Special Theory) were "in the air." He made a great creative leap, but the answers were readily accepted by the physics community. The mathematical transformations that he derived are known as the FitzGerald–Lorentz transformations because FitzGerald and Lorentz had independently proposed these formulae on empirical grounds. The concept of the photon, however, was truly new and not accepted by the physics community for many years.

13 The Creativity Continues: High Mobility Polymer Thin Film Transistors

The market growth for Plastic Electronics is accelerating as the performance of the various devices improves. The transistor is, of course, the circuit element that enabled modern electronics. The Field Effect Transistor (FET), alternatively known as the Thin Film Transistor (TFT), is an electronic switch. The mobility (μ) of the electrons passing through the FET defines how much current can be switched by the FET and how fast that current can be turned on or off. Mobility is therefore the "Figure of Merit" for utility of polymer-based FETs in applications. Although FETs fabricated from semiconducting polymers were first demonstrated more than 20 years ago, μ was far too small to be useful in technology. Over many years, the mobilities gradually increased to values approaching that of amorphous Si (~ 1 cm^2/Vs). To have impact on electronics technology, however, plastic transistors must have mobilities significantly higher; for example, high enough for use in radio frequency identification devices (RFIDs), or high enough to drive the pixels in a hand-held display or a TV — either for high-end LCD displays or for Active Matrix Organic LED Displays. Many thought that this was a hopeless quest.

Only recently have the mobilities of the charge carriers in TFTs fabricated with semiconducting polymers as the electro-active materials been increased to values that will enable semiconducting-polymer-based TFTs to have significant impact. In 2013, in my laboratory, TFTs fabricated from aligned semiconducting polymers were demonstrated with mobility of approximately 25 cm^2/Vs, a value sufficiently large to function in the backplanes for liquid crystal displays and even in the backplanes for organic LED (OLED) displays. Since the display industry is a $100,000,000 per year business, this was indeed important progress.

We have known for years that the optimum transport occurs when the electrons move along the conjugated backbone. The self-assembly of macroscopically aligned polymer chains for use in FETs was accomplished by nano-structuring the substrate upon which the devices were fabricated. In fact, the process was rather simple. At my suggestion, my post-doc, Mike (Hsian-Rong) Tseng carefully scratched the substrate using a kind of "sandpaper" composed of 200 nm diamond particles rather than sand. The "scratching" created nano-grooves in the SiO_2 substrate. The results were spectacular — when deposited from solution, the polymer chains were oriented and highly aligned within the tiny nano-grooves.

Early in 2014, we made another breakthrough leading to better orientation and alignment and a measured mobility of $\mu = 70$ cm²/Vs, a value that I thought that I would never see during my lifetime! This most recent discovery revealed a general and effective strategy for realizing unidirectional alignment and efficient charge transport for semiconducting polymer films deposited on textured (nano-grooved)

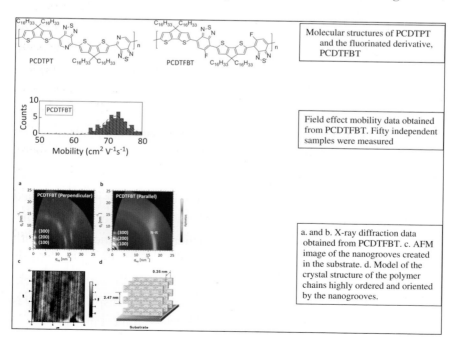

	Molecular structures of PCDTPT and the fluorinated derivative, PCDTFBT
	Field effect mobility data obtained from PCDTFBT. Fifty independent samples were measured
	a. and b. X-ray diffraction data obtained from PCDTFBT. c. AFM image of the nanogrooves created in the substrate. d. Model of the crystal structure of the polymer chains highly ordered and oriented by the nanogrooves.

substrates. By employing a clever sandwich casting method in a tilted tunnel configuration, Dr. Chan Luo was able to utilize capillary action to direct semiconducting polymer self-assembly along the uniaxial nano-grooves on the substrate. The strength of capillary action can be tailored by different surface treatments and functionalization of the spacers mounted at the end of the tunnel structure. Charge transport in the polymer films prepared by this method is sensitive to the strength of capillary action induced by the functionalized spacers.

Semiconducting polymers with different molecular structures may yield even higher intrinsic values. Hence the need for continuing directed synthesis. This methodology has clear potential to be applied to a broader range of semiconducting polymers for diverse optoelectronic applications. The concept of capillarity-mediated self-assembly and alignment opens the possibility of enhancing anisotropic charge transport to create high mobility solution processable FETs for low-cost organic electronics. Moreover, it is quite clear from the fundamentals of the materials science that even higher values will be possible with better alignment of specifically designed semiconducting polymer chains.

Organic LED (OLED) displays require the control of higher currents than the Liquid Crystal Displays (LCDs) that one finds today in television sets, laptop computers, "iPads" and cellphones. However, OLED displays provide the viewer with a superior picture and — it is expected — at a lower cost when the technology reaches maturity. OLED displays are already on the market today. For example, the Samsung "Galaxy" series of cellphones are taking an ever growing fraction of the market share. Some industry experts anticipate that within a few years, OLED displays will dominate the $100 billion display market from mobile phones to computer displays and electronic tablets (like the "iPad") to large TVs. Our recent demonstration of mobility equal to 70 cm^2/V-s implies that printable semiconducting polymer FETs should be able to significantly reduce the fabrication cost of such high-end displays.

The Plastic Electronics Revolution is underway, and major growth in the marketplace is anticipated as scientists and engineers continue to invent new applications and to improve the properties of the semiconducting polymer materials; for example, to continue to improve the mobilities of the semiconducting polymers used in thin film transistors.

"Plastic Electronics" includes a wide range of applications that are based on solution processing of semiconducting polymers. Because these electronic polymers can be dissolved in common solvents, they can be formulated as inks — "inks with electronic functionality" — for use in manufacturing by printing technology. Gutenberg invented printing in 1545. Today, scientists and engineers in laboratories all over the world are printing electronic circuits, they are printing light-emitting displays, they are printing lasers, they are printing photodetectors, they are printing electronic sensors and they are printing solar cells.

The scope of Plastic Electronics currently includes a variety of products, ranging from those listed above to printed batteries and Supercapacitors. Current estimates of the market volume range from a billion US dollars in 2012 to about 5 billion US dollars in 2015. Thin film products on foil or on plastic substrates, manufactured by roll-to-roll processing, will enable the integration of heterogeneous functionalities (displays, sensors, logic and memory) in one mechanically flexible product.

The "Plastic Electronics" Revolution is indeed underway.

14 Decision to Go West – Was it the KGB?

The red, heart-shaped bathtub in the middle of the hotel room in Iowa City, Iowa, saved our marriage.

We departed from Philadelphia bound for a new life in Santa Barbara on a hot day in early August 1982. We went immediately from the law offices where we had signed the papers that sealed the sale of our house on Anton Road in Wynnewood, a suburb of Philadelphia in the area known as the Main Line because of proximity to the main line of the Pennsylvania Railroad. It was a beautiful house, stone with white trim, in the middle of a large property with a stream running through. We even had trout in the stream, although I never succeeded in catching one, I often saw them in the clear water. During the summer, the trees that surrounded the property on all sides gave us the impression of being shielded from the world by a "green wall." In the winter, the snow on the evergreens and on the lawn that sloped gently down toward the creek turned the property into a wonderland. The house had a circular staircase that ascended from the large entry hallway. The dining room was elegant, the living room was 40 feet in length with a view out to the swimming pool. Ruth loved that house — she still does.

We had lived in the Main Line suburbs of Philadelphia for 20 years. We raised our two sons there, and we had many friends there; in truth, we had made a life there, and neither Ruth nor I had contemplated leaving. These ties to the area gave us pause when considering the opportunity to leave the University of Pennsylvania to move across the country to Santa Barbar, where I would become Professor of Physics at the University of California campus at Santa Barbara, UC Santa

Barbara, but we had decided to make the cross-country move. Nevertheless, after having made the decision, the act of actually selling our dream house was a traumatic experience.

Leaving our home in the Philadelphia suburbs was made even more difficult because we had experienced a wonderful summer. Our son Peter and his fiancée, Shauna, were married on June 12, 1982. In many ways, that wedding was the culmination of our 20 years in the Philadelphia area. We invited family from all over the country, friends from our life in the Philadelphia area and colleagues from Penn. Even today, 32 years later, when we see these friends, colleagues and family members, that wedding is always fondly remembered as "the best wedding ever."

Shauna's family was from the New York area; her father, Bennett Bakst, owned a business on the corner of 50th and Lexington, an upscale drugstore stocked with everything that one might expect from a drugstore in Midtown Manhattan, including expensive cosmetics, a prescription pharmacy and a lunch counter. The drugstore was open 24 hours a day, seven days a week. Famous people from the New York scene were regular customers. On cold winter nights, the lunch counter was a warm place that drew diverse customers, including the "working girls" from the surrounding area.

The weekend of the wedding was a major New York experience. Through Bennett's connections, Ruth and I rented the penthouse apartment in the San Carlos Hotel, where we royally entertained our guests with drinks and a spectacular view of the Manhattan skyline.

The weekend started with a prenuptial dinner on Friday evening. Ruth and I hosted the dinner at Sammy's Rumanian Steak House? on the Lower East Side. We reserved the entire restaurant. Sammy served traditional Eastern European Jewish style foods, rich and full of cholesterol. When we made the arrangements, Ruth asked about salads. Sammy responded that Jews do not eat "green." Needless to say, we had no salads. Instead, we had chopped liver, matzoh ball soup, schmaltz served from pitchers on the table, and huge veal chops that were too large to fit on the plate. We had arranged for a group to play music and sing traditional songs for the guests. Because this group participated in the service at a nearby synagogue, and because the service could not start until sundown on one of the longest days of the year, the dinner went on until after 11 pm. When we departed, we saw a long line of customers waiting for tables.

Because of a now-famous anti-nuclear demonstration in Central Park on June 12, 1982, there were literally millions of people on the streets of Manhattan that Saturday. But the main event was indeed the wedding at a synagogue near the Bakst home on Long Island. Again, because of the late sunset, the wedding affair did not even start until 9 pm, and it was nearly midnight before the actual ceremony took place. The Baksts hosted a classic Long Island Jewish wedding, sumptuous and generous, the kind of affair that you might see mimicked in a Hollywood film. The evening started with drinks and appetizers served at a series of stations strategically placed around the large ballroom. Before the ceremony started, I had plenty of time to celebrate — and I did. I walked down the aisle well-supported by Ruth, who held tightly onto my arm. It was a magnificent wedding with dinner following the ceremony. We returned to our penthouse in the San Carlos at nearly 5 am with dawn showing in the eastern sky.

After the wedding, we began the job of packing all of the many things we had accumulated in the 20 years of our life in the Philadelphia area. Simultaneously, I supervised and participated in the packing of the equipment from my laboratories in the Laboratory for Research on the Structure of Matter at Penn. I was determined to "hit the deck running"; I wanted to have my new labs functioning at UC Santa Barbara without significant delay.

The final settlement of the sale of a house in Pennsylvania is very real and personal. In many parts of the country, the details of the sale are accomplished "remotely"; the deal is made through the realtors as intermediates, funds are placed in escrow, and the final papers are signed impersonally without a need for the parties to physically meet.

In Pennsylvania, however, the buyers and the sellers were required to sit around a common table together with lawyers and real estate agents. The actual legal documents defining the sale and the transfer of funds were passed around and each was signed by both parties. In that office in early August 1982, this signing ceremony was, to Ruth and to me, the end of a significant chapter in our lives. The house on Anton Road was a symbol to us of the long journey starting with our departure from Omaha, newly married and with a total of $618 in our treasury, to our life in the academic world of a great university — and it truly was a beautiful house.

We walked out of the offices where the signing of the documents took place, got into our car that had been previously loaded with the baggage needed for the trip, and pointed the car onto Interstate 76 West. We were in the car for only a few minutes when a major argument broke out. I do not remember the subject about which we argued because the subject was irrelevant; for the "fight" was really about the selling of our house and the risks and uncertainties that are implied in such a major move.

I was driving and Ruth sat in the passenger seat. We were both very angry, and we stopped speaking to each other. We traveled across Pennsylvania and spent that night in a hotel near Pittsburgh. We did not speak while in the car, and we did not speak that evening during dinner at a restaurant near the hotel. We did not speak to one another the next day as we traveled through Pennsylvania, Ohio and Indiana, through the complex pattern of tollways that surround Chicago, and through the rest of Illinois. We were silent as we crossed the Mississippi River into Iowa. As that long, long day came to a close, we found ourselves on the highway outside of Iowa City. We stopped at two motels only to be told that there were no vacancies, for there were many families on the road traveling to or from summer vacations. We then pulled into the Knights Inn. When I asked if there were any rooms available, the clerk at the registration said they had only one vacancy, but he insisted, with a somewhat strange smile on his face, that I inspect the room before I signed the register.

The red, heart-shaped bathtub in the center of the room and the round bed with the mirror on the ceiling above saved our marriage. When we walked in to inspect the room, we both broke out laughing. I ran back to the office, signed the registration, and hurried back to join Ruth in the red heart-shaped bathtub. We were on our way to our new life; we had plenty to talk about and much to look forward to in this start of our adventure in Santa Barbara.

I had been at the University of Pennsylvania for 20 years, moving up the academic ladder from Assistant Professor to Associate Professor and then to Professor. I had built an international reputation and had initiated the pioneering research that would lead eventually to the Nobel Prize. I had served for six years (1974–1980) as Director of the interdisciplinary materials research laboratory at Penn, the Laboratory for Research on the Structure of Matter. I had even served in a senior position within the university administration, as Vice Provost for Research.

I knew Penn inside and out, and I was proud to be a part of this venerable and respected major research university. Now we were leaving and on our way to UCSB, a set of initials which in 1982 stood for University of California "Sunny Beach." As one of the youngest campuses of the University of California, UCSB was a relative newcomer to the academic scene and certainly did not have stature in the academic world. How and why had this happened?

A group of four relatively young theoretical physicists, Douglas Scalapino, James Hartle, Robert Sugar and Ray Sawyer, known locally at UC Santa Barbara as the "Gang of Four," had successfully competed for support from the National Science Foundation to create the Institute for Theoretical Physics (ITP). This was to become a national institute. The founding of the ITP was a major success for physics at UC Santa Barbara and a turning point for the Santa Barbara campus. The Gang of Four together with the Chancellor, Robert Huttenback, quickly recruited Bob Schrieffer and convinced him to leave Penn to join the ITP. My future colleague and future Nobel Laureate, Walter Kohn, was recruited from

UC San Diego as the founding Director of the ITP. Schrieffer became Director after Kohn.

As soon as Schrieffer arrived in Santa Barbara, he began recruiting me in subtle and not so subtle ways. He would call me on a wintry day in Philadelphia and taunt me with the beautiful sunny weather in Santa Barbara. While speaking to me, he would describe the dolphins jumping (from his office, he had a spectacular view of the ocean) and the whale that he had seen breaching earlier in the day. In the 32 years since I first arrived in Santa Barbara, I have actually seen dolphins jumping, but never have I seen a whale from the beach.

He also recruited me with discussions of the exciting physics that he was doing on the generalization of the theory of solitons in polyacetylene, the subject on which he and I had collaborated during the two years before he left Penn.

In the spring of 1980, the Institute for Theoretical Physics organized a Symposium on One-Dimensional Conductors and the Insulator-to-Metal transition in Conducting Polymers. I was invited to give the opening lecture. Schrieffer and Jose Fulco, then Chairman of the Physics Department, greeted me at the Santa Barbara airport. Even before we departed from the airport toward the nearby UCSB campus, they informed me that they had arranged a meeting with the Chancellor later in the day, and that I should think seriously about what would be required (resources, space, salary, etc.) to bring me to Santa Barbara.

At that meeting, I told Chancellor Huttenback that the promise of an Institute devoted to my interests in semiconducting and metallic polymers might be the foundation for an offer that I could not refuse. Promises were made "on the spot." The situation was further complicated the following morning. I went out of the hotel for an early morning run, turned the corner and stopped at the sight of a beautiful Triumph Herald, precisely identical to the one that we had purchased during our sabbatical leave in England in 1964. Although the fenders and floor of my Triumph Herald had rusted through and the left front wheel had come off while driving on the Schuylkill Freeway years before, the Triumph

Herald parked by the side of the road in Santa Barbara in 1980 looked as though it had been driven from the new car dealer that very same day. If Santa Barbara could be that good for a Triumph Herald, just think what life in Santa Barbara could do for me!

Ruth had accompanied me on the trip; we were wined and dined and treated royally. This was clearly going too fast — I was being swept off my feet. Nevertheless, as a result of the events that occurred during the three-day symposium, we departed from Santa Barbara thinking seriously about going west.

Reality returned when we arrived home in Philadelphia. My science was going well, I had recently accepted the appointment of Vice Provost for Research (on an "acting" basis, since neither I nor the university had committed to this possible alternative future), the Penn campus looked great in the bloom of spring, and my colleagues in the physics department at Penn almost immediately responded with a pre-emptive salary increase and funding for some new equipment that I needed for my research.

My sojourn into university administration was a symptom of a deeper problem. Truth be told, I was 45 years old and going through a midlife crisis. I was not optimistic about growing old doing physics. Would I be able to continue to take risks with new ideas? Where would the new ideas come from? Would I really be able to remain creative? Despite the current progress and the excitement generated by new results from my laboratory, the transition into university administration seemed easier and perhaps safer.

As a result, on a snowy evening in January of 1981, I met Bob Schrieffer in Princeton, where he was visiting for a few days, with the intent of saying "No" to the offer to come to UC Santa Barbara. We walked across the Princeton campus with large snowflakes falling gently around us. We had drinks and dinner and by the end of the evening, it was done. I informed him that I had decided to stay at Penn.

Two events subsequently convinced me that, indeed, science was my life and that I should reconsider the possibility of moving to Santa Barbara.

The first of these two events occurred only about one month later. I was invited to a conference held in a hotel near Daytona Beach Florida. This was one of a series of well-known annual conferences, held initially and for many years on Sanibel Island (Florida), and therefore continuing to be known as the Sanibel Conferences even after they moved to Daytona Beach for financial reasons. I gave a lecture on our recent results. The conference schedule was relaxed, with lectures in the morning and in the evening after dinner, but the afternoons were free of any scheduled events. The idea was to leave time for discussion and interaction.

Professor Kenichi Fukui participated in this Sanibel Conference. He was a distinguished and respected theoretical chemist from Kyoto University in Japan. Professor Fukui was awarded the Nobel Prize in Chemistry later that same year (1981); he shared the Prize with Roald Hoffmann of Cornell University for their work in theoretical chemistry. During one of the relaxed afternoons, I was sitting quietly and talking with Fukui-san — but I didn't really know who he was. He was very much a gentleman, and he knew me better than I knew him. He was well aware of our breakthrough research on conducting polymers.

During our relaxed discussion, he asked me what I was doing, and I openly expressed to him my uncertainty about my career plans. I told him that I was serving as Vice Chancellor for Research and considering a career change in the direction of university administration. Why would I say that? He was not a close friend; he was almost a complete stranger! In a characteristically quiet but forceful way, he stated unequivocally that this would be a mistake, that science was more important and that I should dedicate myself completely to my scientific endeavors. That I remember this discussion with such clarity demonstrates that the advice had impact on my career and my life.

The second event also involved a scientific conference. In June 1981, the first of what has become a series of conferences, the International Conference on Synthetic Metals (ICSM), was held in Boulder, Colorado. The five days of that first ICSM were filled with excitement,

excitement for this new scientific field, excitement with the new data that were introduced in almost every lecture and excitement generated by the controversies that were generated around the interpretation of these new results.

I came to this conference with a briefcase filled with new data from my labs and a stack of preprints on the latest results from others. In 1981, there were no laptop computers; I carried raw data, typically in the form of graphs printed directly off the x-y recorder during the experiment or graphs made by hand summarizing the data. I had no copies of these graphs.

Every day during the five days in Boulder, I made two telephone calls to the post-docs and students in my labs to find out what new results had been obtained, to give them advice on what to do next, and to inform them of events at the conference. On Friday, the final day of the conference, I realized that I had not called my assistant at the Vice Provost's office even once. The message was clear to me. Professor Fukui had been correct: Science was indeed more important to me, and I should therefore dedicate myself completely to my scientific endeavors.

I called my friends in Santa Barbara the same Friday afternoon and told them that I wanted to re-open the discussions. We agreed that I would come to Santa Barbara the following week after participating in a second three-day conference at Stanford University in Palo Alto.

The conference at Stanford involved a small group of scientists, among whom was Fred Wudl. Fred was on the scientific staff of the famous Bell Laboratories and was well known as one of the most creative and productive synthetic chemists on the newly created field of organic conductors. Fred had also been approached by UCSB. The two of us, together, could (and subsequently did) create a program that would be second to none in the world. We talked endlessly about this opportunity, and without any overt statements, we both decided during those three days at the Stanford Conference that we should make the offers by UCSB sufficiently attractive that we would move to Santa Barbara to set up the Institute for Polymers and Organic Solids. The three days at Stanford prepared me for the visit to Santa Barbara, where I would make the deal that resulted in our move to UCSB.

Ruth accompanied me to the conference in Boulder, and we traveled together to Palo Alto and subsequently to Santa Barbara. In Palo Alto, we checked in at the Holiday Inn immediately across El Camino Real from the palm-tree-lined entrance to Stanford University. Because we had lived in Palo Alto for one year in 1958–1959 when I worked in the Lockheed Missiles and Space Division while continuing my graduate studies part time at UC Berkeley, we looked forward to revisiting our favorite places. At the end of the final day of the Stanford Conference, we went out to dinner at a simple restaurant that we had enjoyed many times. We were not gone long, perhaps an hour and a half. When we returned to our room on the first floor of the Holiday Inn, Ruth immediately said: "Someone was in here!" Some item that she had left on the dressing table had been moved. Initially, I denied the possibility because there was no indication that we had been robbed. Then, I realized that my briefcase was gone!

My briefcase was gone; all the data were gone, all my notes from the Boulder Conference were gone. I was devastated. I called the police and searched the area around the hotel with the two officers who responded to my call. They suggested that the burglar might take anything of value (such as traveler's checks and airline tickets) and then discard the briefcase. But alas, we did not find anything. I informed the airline, and our tickets were reissued on the following day. There was nothing else in the bag of negotiable value, no money, no credit cards, and no traveler's checks. But all the data were gone and all my notes from the Boulder Conference were gone. To me this was a disaster. The data were precious and irreplaceable. There were no computers at that time, no way to store data except as numbers or graphs on chart paper. Of course one can always say that the experiments could be done again. However, there were many, many hours involved in acquiring these data, especially the specific electron magnetic resonance data that demonstrated the existence of a first-order phase transition from semiconductor to metal.

Being forced to recreate these data would set me back many months. Remember the importance of "Who did what first and was it right?"

Nevertheless, we went on from Palo Alto to Santa Barbara, and that final stop on this long trip was successful. We were able to come to agreement on the details of the move to UCSB. Although several months were required to finalize the offer, the university was eventually able to deliver on all aspects of the offer that we had discussed during that visit. We departed from Santa Barbara, with anticipation for the future. I would have been elated were it not for the stolen briefcase and the loss of all the original data.

Life is Good, and I am indeed a lucky man. When I went to my office the following Monday morning, I found, squarely in the center of my desk, a large brown envelope that had been in my stolen briefcase. There was no address on the front, but the return address was printed in the upper left hand corner:

Professor Alan Heeger
Laboratory for Research on the Structure of Matter
University of Pennsylvania
33rd & Walnut Streets,
Philadelphia, PA.

In that envelope were all of my cherished notes and data. The "thief" had sorted through everything in my briefcase and placed all of the original graphs and documents (and only the original graphs and documents) in the envelope and then dropped that envelope into a US Mail box. Any document that was clearly a copy (for example, reprints) was not returned; only those documents which were clearly original and without obvious copies available were inserted. The post office had dutifully returned the envelope to the return address with "Insufficient address; Insufficient Postage" stamped in red across the front. This was a miracle; the second miracle in my life!

A number of theories were proposed in discussions of how this could possibly have come about. One possibility is that the thief was not just a "smart thief," but a graduate student at Stanford working his way through school, perhaps a physics graduate student with a sufficiently strong scientific background to know what would be important to me and with sufficient sensitivity to realize how serious the loss would be to me. This explanation makes no sense, however, because there was nothing of value taken from the briefcase. The airplane tickets were in our names and could not be used by anyone else. They had been quickly replaced.

My favorite theory, however, was that I was robbed by the KGB! This event occurred in 1981, during the Cold War, immediately after two well-publicized conferences in a new field in which I was known to be the leading player. The KGB theory was bolstered by the fact that nothing in the room was taken other than my briefcase. Ruth's jewelry was not touched, and the bags containing our clothing and other possible valuables had not even been opened. It seemed that the sole purpose of the break-in had been to take my briefcase. Whether it was a KGB agent, I will never know. The good news is that those precious data were subsequently published in some of my favorite scientific papers. Included, as noted above, were the data that established the sharp transition to a metallic state in polyacetylene as a function of the doping level.

15 Chapter

Life and Science in Santa Barbara

After leaving Iowa City, the remainder of the drive across the country to Santa Barbara was enjoyable and relaxing. We drove through Colorado, visiting Durango and Mesa Verde, and then through Arizona, where we stopped for a few days at the Grand Canyon. We arrived in Santa Barbara on August 10, 1982. Our furniture arrived on August 11, 1982, on the 25th anniversary of our wedding day, and we moved into our Santa Barbara house on the Riviera high above the city with spectacular views of both the sea and the mountains.

After the decision to move had been made in early 1982, Ruth and I met in Santa Barbara; she came from Philadelphia, and I was returning from a trip to Asia. Our purpose was to find a house. We woke up early on the Sunday morning following our arrival, bought a local newspaper, walked down the block to a coffee shop, and we went through the real estate listings while having a light breakfast. Since we did not yet know the city at all, we circled a few entries that seemed to have promise, houses that we thought we might be able to afford and houses that were situated on the Riviera where we knew the views were spectacular.

Because we had little else to do that morning, we purchased a city map and drove around the area. The first house we saw was the one we eventually purchased. We had a real estate agent (actually a relative, the sister of my cousin), and we asked her to show us this particular house first. I was immediately ready to make the decision. Ruth needed to see 30 other possibilities, but finally came around to my point of view, and we bought the house at 1042 Las Alturas Road. Our new house was situated approximately 950 feet above sea level, with spectacular views of both the Pacific Ocean and the mountains.

As expected, the movers unpacked and helped to put the new house in order. Of course, a thorough cleaning was required and some initial touch-up painting of the walls where the previous owners had hung their pictures and objets d'art. I then immediately began (with Ruth's permission) to hang our meager art collection, and then went off to the University of California at Santa Barbara to start the process of putting my lab back together. We were re-settled (or at least re-settled to the point where we could begin build our new life in Santa Barbara) in a remarkably short time. Fortunately, we had friends, former colleagues from Penn and their families, who made the transition somewhat easier.

We completely remodeled and enlarged the house a few years later. Today, we have our walls covered with paintings, many of which are especially interesting and colorful and were done by Ruth. She has a natural talent for painting with oils, and she had taken instruction from professionals on and off over the years. Her paintings are displayed with pride all over our house.

We also were able to realize a long-term dream of mine — to have a sculpture garden. The sculpture garden began with a trip to Berkeley, where I participated in a scientific conference. We took off on Sunday afternoon, crossed the Bay Bridge and drove north toward Bodega, CA, the town that was made famous by Alfred Hitchcock's movie, *The Birds*. As we approached Bodega, we saw an open field on our left with many redwood figures, some very primitive and unfinished and others that were quite interesting. The sculptor, Richard Stoppard, was shirtless and working on a tall female figure that immediately drew our attention. We stopped and talked to him at some length. He told us that her name was "Fiona," but that she was not yet finished. His style was to leave the redwood rough and unpolished, so we were to expect her in that form. She is tall, with prominent breasts, a haughty smile — truly a beauty. We immediately loved Fiona, agreed on a price, and made a deal to purchase her. He said that he would be finished within a week or two at most and that he would send her to us by Federal Express. We walked around a bit, and visited the schoolhouse where the dramatic scenes from the Hitchcock movie were filmed. Near the entrance stood a small Indian crouched down in a sitting position. Although the wood had cracked, he was nevertheless interesting and somehow special. Richard gave the squatting Indian to us free of charge. It, too, now stands in our sculpture garden.

Approximately two weeks later we received a call from Richard. He said that he was not able to put Fiona in a box; it would be too much like putting her in a coffin! Therefore, he proposed to drive to Santa Barbara with Fiona in the back of his station wagon — and he did. We had already decided where to put her. The area in front of our house is surrounded by a relatively high white wall that would protect her. She stands in that protected area with her body oriented toward the sea, but her face is turned toward the gate where we and our guests enter. We attached her to a metal disc which was screwed deeply into the ground and covered the disc with the small stones that covered the entire area. She stands there today having never been moved, and she still looks haughty and beautiful. For a few years, I carefully oiled her with a wood treatment that would keep the redwood looking fresh and new. This was basically an erotic experience for me. Eventually I stopped, so her color has evolved to a lovely brownish-red-grey. Fiona was the first purchase for our sculpture garden.

Richard began to visit us fairly regularly, always short of money and always with a sculpture in the back of his station wagon. But he was fun, and we greatly enjoyed his visits, and enjoyed taking him out to dinner. More importantly, we enjoyed the purchases we bought from him. We bought his full size image of Senator "Dianne Feinstein" and placed her in back of the house near the edge of the swimming pool. She, too, is still there looking dominating and competent. He made for us a small bear that was shaped so that it looked as if it were climbing a tree — and we attached it to a tree in a very realistic way. When my five-year-old grandson visited a few months later, he said: "Look Dad, there is a bear on the tree out there!"

The next purchase was a half-propeller from an early single-engine airplane. It was properly mounted on a red base and had beautiful curvature. It was a propeller, but it was clearly also a work of art. And so it went on. Ruth found or created most of the art and, typically, after discussion with me, we made the purchase. We brought pieces back with us from trips abroad. We acquired a figure with a beautiful abstract shape that has the topology of a donut (with a hole in the middle), the color of which matches perfectly with the small stones that cover the space. We installed a steel fountain in the form of several flowers made by a student of Aris Demetrios.

Aris is an accomplished sculptor and a good friend, with pieces installed (many very large) all over the world. He made for us an abstract brass-colored cube with an involuted internal structure that is mounted on a base on one of the cube corners. This abstract cube catches the light at almost any time of day. It stands directly in front of the entry to the house. More recently, we added an attractive Blue Dog, later a Blue Cat and some stone carvings. Because the sphere is to me a symbol of perfection, there are spheres of various sizes distributed in patterns throughout the space, copper spheres and stone spheres. Now we have the sculpture garden that I had dreamed of — and it is quite special.

Inside the house we have a collections masks — masks from Asia (Japan, Korea and China), masks from Venice, and a porcelain face of a woman that we found in Alsace. In Taos, we acquired a large ceramic piece built around the face of a woman. When I am awake in the middle of the night, I talk to her. In Xi'an, we were fortunate to find an ancient (more than 300 years old) helmet/face made of brass that now has a green patina. Thus, inside the house, there is always a group of faces looking at me.

Not long after we had settled into our new life in Santa Barbara, we traveled to Greece for a summer school focused on the current status and the future promise of the growing field of semiconducting and metallic polymers. We started the trip by spending a week in Athens, where we enjoyed the antiquities, we enjoyed the food, and we enjoyed

the Greek people that we met in the shops and restaurants. We had great weather; no sign of the pollution that they often suffer in Athens.

We left Athens on a cruise ship that carried us on a tour of the Western Mediterranean and visited many of the beautiful Greek islands. We went even as far as Ephesus (Turkey) before returning to Athens. Upon our return, we immediately embarked on a small, but very fast, hydroplane that took us to the destination island for the summer school; we remained there for three weeks in a lovely hotel a few blocks from the central shopping area and the port.

While shopping one day, Ruth chose three black shirts of the kind seen everywhere in Greece as souvenirs; one for me and one for each of our two sons. She had bargained with the proprietor and come to agreement at a price roughly half what he had quoted originally. I independently chose the same three shirts while strolling through the shop and came up to the counter with the intent of purchasing them at full price! This caused quite a lot of laughter and proved, once again, that I am a lousy negotiator. We bought the shirts at the price Ruth had negotiated, and I began to wear mine. The black looked good on me, with my white hair, in the informal atmosphere of the summer school.

As a professor at Penn, I always wore brown. My hair was brownish-red, and brown tweeds seemed the appropriate dress at the University of Pennsylvania, a traditional Ivy League university. I was so well known for wearing brown clothes that the wearing of brown was used to describe me to strangers. For example, once, when visiting Japan, I was waiting to be met at my hotel by a Japanese scientist who worked at the company I was scheduled to visit. I sat in the lobby for a relatively long time, well beyond the time previously agreed. I noted that there was only one other person sitting quietly in the lobby. I approached this Japanese gentleman, and asked if he were waiting for Alan Heeger. He said: "Yes, but I was told that you always wore brown."

I now wear only black. After returning to Santa Barbara, I awoke one morning and noticed again when I looked in the mirror that I had white hair. I owned the one black shirt that I had purchased in Greece, but I made the decision — and a rapid and complete transformation occurred. I now wear only back; even my underwear is black! I need not even turn on the light in my closet, for every item of clothing is black. This transformation was not a direct result of the move to Santa Barbara. It was, however, consistent with the beginning of our new way of life.

When I subsequently became seriously interested in the conversion of sunlight into electricity and started my quest to make efficient solar cells fabricated from semiconducting polymers, the "Man in Black" image was even more appropriate. Black implies that all the colors of the spectrum are absorbed, precisely what one wants for efficient conversion of sunlight into electricity.

These days, because I only own black clothes, there is never any ambiguity when I am scheduled to meet a stranger.

Santa Barbara is a gem, a city of approximately 150,000 inhabitants right on the edge of the Pacific Ocean (approximately 200,000 if one includes Montecito and Goleta). When you look at the map of California, you will notice that the coastline turns east just before reaching Santa

The Man in black.

Barbara from the north. Thus, when I look out at the sea from the deck of my house, I am looking south. At certain times of the year, I can watch both the sun rise and the sun set over the Pacific Ocean. In December, when I look sleepily in my bathroom mirror early in the morning after waking up, the sun is reflecting directly into my face. Even after more than 30 years, I find this somewhat confusing and disconcerting. We live on the west coast, but in the summer, the sun rises and sets over sea. In the winter the sun rises directly from the sea, and the sunshine penetrates deeply into the house.

Santa Barbara is a one-dimensional city running along the coast and pinned to the coast by the mountains behind. Because I have been involved in the science of one-dimensional materials (such as long chain conducting polymers) for so many years, it is the perfect place for me. The mountains behind Santa Barbara trap the moisture from the ocean (morning fog that leaves the ground and plants covered with dew). As a result, when looking at Santa Barbara from a boat out beyond the harbor, one sees a "green city" rather than the famous California Gold, which in my opinion is simply dried out brown grass.

For a city of about 200,000, the cultural opportunities are remarkable. All the great musicians and orchestras that come to the West Coast (to Los Angeles or to San Francisco) stop in Santa Barbara. We have heard Yo-Yo Ma at least five times in recent years. Joshua Bell and Yitzhak Perelman were on the 2013–2014 season. The Santa Barbara Opera, CAMA, the UCSB Arts and Lectures program (which often has performances in the theaters in downtown Santa Barbara), the Music Academy of the West and the State Street Ballet broaden the cultural spectrum.

The Ensemble Theatre Company (ETC) has been putting on plays for more than 30 years. When we first arrived, it was indeed an "ensemble" theater; the actors and artistic director were all locals. Nevertheless, because of the close proximity to Los Angeles, there are many talented actors in Santa Barbara. Even in the early years after our arrival, I recall a performance of Arthur Miller's *Death of a Salesman* that left me breathless at the end wondering how it would be possible to do such a performance again and again. We first experienced *Master Class* in that early period, a play that I mentioned earlier because of the line that affected me so deeply: "I did not lose my voice — I lost my nerve."

Over the years, the ETC has evolved to a professional regional theater with a talented artistic director, Jonathan Fox, who draws upon professional actors from New York and Los Angeles. A new theater was completed in November 2013 to complete the transformation.

Now back to science, science at UC Santa Barbara: The moving trucks that brought our furniture and all that had been in our house in Philadelphia also carried all the equipment from my laboratory at the University of Pennsylvania. I had convinced nearly all my research group, graduate students and post-doctoral researchers, to come on this adventure to California with me, and during the month prior to our departure, we carefully packed up the lab into boxes that would provide protection from damage by the usual methods of packaging with paper and bubble wrap. The large equipment was moved "as is," but attached to wooden pallets to prevent motion and damage. Our goal was to "hit the deck running"; we hoped to get the new lab up and operating within a few short weeks and to be able to begin acquiring new data within a month — we actually succeeded in doing so. In addition, as part of the incentive to move to UC Santa Barbara, I was given a "funding package," funding for new equipment that I needed to start research in new directions. This equipment had been ordered before we departed from Philadelphia, but setting it up required additional time.

When I arrived at UC Santa Barbara in the summer of 1982, I concluded that I needed to have as part of my lab everything that would be needed for our research. I wanted to be independent, and I designed the lab with that in mind. Today, the situation is very different. The Materials Department at UCSB, in particular, has a very open structure. In most universities, professors have their own "turf," but not at UC Santa Barbara; at UCSB interdisciplinary collaboration is the mode of operation. There are many central facilities available for all to use. In recent years, my students bring me data that they have acquired using equipment that belongs to other professors. Not infrequently, they work with and publish with one of my colleagues without any involvement from me. Our research atmosphere has become very transparent and is known and admired for this interdisciplinarity and transparency.

Prior to the move, one of my concerns was the question of whether anyone would actually be able to work. In Philadelphia on a rare sunny day, the graduate students would tend to take off and enjoy the day rather than focusing (as I thought they should) on their thesis research. In Santa Barbara, nearly every day is beautiful with blue skies and moderate temperatures. On a typical day, if one looks straight up with a pair of polarized sun glasses, one sees that the incident light is polarized, implying only a single scattering event rather than the multiple scattering that occurs when light passes through an atmosphere containing a relatively high density of particles (from industrial plants or automobiles). In Santa Barbara, the air is clear.

The University of California at Santa Barbara is located on a peninsula approximately 10 miles north of the center of the city; it is a beautiful campus where the physics buildings is at most a few hundred yards from the sea. On the UCSB campus, the girls are beautiful and in the manner of California girls, they seem to wear very little. Would anybody work?

In fact, my worries were unfounded. Everyone dug in and seemed to work as hard as graduate students everywhere. As a result, we have been remarkably productive over the 30 years in Santa Barbara. Nearly 1,000 publications have appeared in internationally recognized scientific journals. I have been blessed to have bright and talented graduate students and post-docs, one after the other, join the group and make important contributions.

The ethnic origins of my students and post-doctoral researchers, however, have changed significantly over the past 30 years. When I departed from the University of Pennsylvania, the members of my research group were mostly American born, with one Japanese post-doc. Today, the numbers are very different. I continue to have the benefit of working with very bright young American students and post-docs from some of the best universities in the United States, but the majority of the post-docs are now from Korea and China. The science in Japan has developed to such a high level that there is no need for a young Japanese scientist to come to the United States for a post-doctoral period. I am pleased to report that my Korean and Chinese post-docs are creative, innovative and productive. I have not found the criticism that the Asian education suppresses creativity to be true. Perhaps we attract those special students who are not spoiled by the rigidity of the education in the Asian countries, or perhaps their educational system has evolved to a new level where these qualities are encouraged and rewarded. For example, Dr. Chan Luo, who succeeded in achieving the high mobility transistors described in Chapter 13, made a major creative step with his use of "capillary action" to self-assemble the macroscopically oriented polymer chains. The long list of such creative Asian post-docs who worked with me and made similarly original contributions can easily be seen by looking at the authorship on many of my most import (and most highly cited) publications.

We quickly began to build up a list of new publications that, once again, exerted leadership in the field of semiconducting and metallic polymers. Our focus in the 1980s was on the physics and chemistry of soluble polymers, for we envisioned manufacturing by use of low-cost printing technology. Based upon the response at scientific meetings and publications in the scientific literature, I was satisfied that we had made the move from Penn to UC Santa Barbara successfully and without a loss in forward momentum.

Even in the paradise of Santa Barbara, there are two ongoing sources of concern and worry: earthquakes and wildfires. All of California sits on or near an earthquake fault. We experience minor earthquakes regularly, by now without concern. There continues to be talk about the "Big One" to come, but so far we have not suffered a serious earthquake.

The Mediterranean climate implies significant rainfall during the winter months from December through March and then no rain at all for the next six or seven months. Thus, every autumn, the back country is dry and ready to burn. There was a big fire in Santa Barbara in 1989 that swept rapidly down one of the canyons from high in the mountains to the sea. More than 500 houses were lost, some belonging to friends of ours. Fortunately, the canyon that guided the fire was approximately five miles north of our house. I went outside that evening and climbed to the top of a nearby hill. One could see the live flames furiously consuming everything in the long path down to the sea. I came in somewhat shocked and said to Ruth that it was inevitable that someday, our house would burn. There is really nothing to do to prevent such a catastrophe. The high winds drive the flames forward at a remarkable speed, and our house is on the edge of a canyon that is susceptible to such wildfires. In fact there was such a fire that passed near our property in 1977 (before we came to Santa Barbara) and left scarred trees but caused no damage to the house.

Our big fire occurred on November 11, 2008. I was on my way home when my cellphone rang. It was Ruth, nearly in panic, telling me that a huge fire was moving in our direction and that we had only a few minutes to evacuate. I heard but did not really believe until I came over the top of a ridge and saw the flames approaching. What does one take out in such a situation? I grabbed a large trash container and ran through the house putting inside all the family photos that I could find. We took a few additional non-replaceable "treasures," packed two small pieces of luggage with the essential clothes and medicines that we would need, quickly tossed these items into our two cars and drove away from the house. As we left our small street, we saw the two houses at the end explode. We met again at a restaurant/bar on the beach and sat watching (while drinking vodka martinis) the progress of the fire through the windows of the restaurant and simultaneously followed the movement of the fire as described by the news on television. It was impossible to identify the path of the fire from so far away, but it seemed quite clear that our house was directly in its path.

We stayed that night with friends. By morning, the fire had burned itself out and the police were putting up barriers to prevent access to the entire Riviera area. I could not wait. I drove to within a few blocks of where the barriers were being constructed, parked my car, and sneaked past the police on foot and hiked up the hill covering the two miles to our house as quickly as possible. As I rose up to the Riviera, I began to see the devastation, entire houses gone with no trace except the ashes. I did not know the fate of our house until I reached the corner of our property. Miraculously, the house was still there, seemingly unharmed. The fire had burned all the plants and flora right up to the foundation. On looking more closely, I saw that many of the roots were still smoking and burning. Fortunately, the water to the house remained connected, I brought out the hose and began to systematically put out these remaining flames. I then noticed that the shrubs in front of our next door neighbor's house were still burning furiously. I put them out and walked around the properties to see that nothing more was burning. Our two houses stood nearly alone. All the houses in the street below us were gone. The two houses at the end of our street were gone. I used the hose to wash the ashes from our deck and to generally start to clean up.

Of course, I had immediately called Ruth with my cellphone to tell her that the house had survived and to give her a brief report of the current status. The word quickly spread that I was up in our neighborhood and people called Ruth to ask her if I would check out their homes. I did so — some were there and some were gone. I reported the good news to those whose houses remained. I refused to give the bad news to anyone; for them, knowing the extent of their disaster would not help.

Considerable time was required to get our place back in order. A thorough cleaning was required. New rugs were needed because of the smoke damage, and all the property needed replanting. Fortunately, we had good insurance coverage, so the out of pocket expense were relatively minor. Nevertheless, it was an experience that I will never forget. We nearly lost everything, but we were remarkably lucky.

Santa Barbara is a gem and a wonderful place to live and to work, but the risk of a wildfire returns every year.

16 The California Entrepreneur: UNIAX Corporation

Every year, Forbes magazine publishes a list of the 400 wealthiest Americans. In 1983, that list was published in the Fall issue of Forbes magazine. My name was not on the list. In a separate, following article in the same issue, Forbes published a list of 100 names which — they predicted — would be included in a future list of the wealthiest Americans, people who were well on their way to a level of financial success that would place them on the Forbes 400 list. My name was not on the "to be" list either.

A third article in that same issue was titled with the provocative question: "If they're so smart, why aren't they rich?" My name was among those featured in the "If they're so smart —" article.

This third article argues that although ingenuity makes the world a richer place, it does not inevitably follow that being "smart" will do much to enrich the ingenious inventor. This statement is of course obvious, but the article is, nevertheless, one of my treasures. I have a framed copy of the first page hanging on the wall of my study. The first paragraph also begins with a list:

> Here are some names you are not likely ever to see in the Forbes Four Hundred: Franklin Lim, Gary Kildall, Bill Gates, John Kemeny, Ananda Chakrabarty and Alan Heeger. What do they all have in common? Each is an American inventor, responsible for a breakthrough technology. Each inventor is still alive and young enough to enjoy the fruits of his labors. All can be said to have done moderately well, a few are even millionaires. Yet for a variety of

196 Never Lose Your Nerve!

reasons, none of them, based on current form, is likely to transcend $125 million in net worth.

The Forbes visionary was obviously incorrect, for today, Bill Gates is considered to be one of the wealthiest men in the world. But in 1983, Bill Gates and Alan Heeger received equal billing from Forbes!

Without doubt, this article had an impact on me. I did not expect to become one of the top 400, but the idea of starting a company to develop our discoveries into applications of commercial value was certainly attractive — and inevitably resulted in my becoming entrepreneurial by creating "start-up" technology companies.

After coming to UC Santa Barbara in the summer of 1982, a few colleagues and I began to build the new Materials Department. The Dean of Engineering, Robert Mehrabian, had acquired the necessary positions through negotiations with the Chancellor as part of the incentive to bring him to UCSB. Because we had not previously had a Materials Department, we had complete freedom on our choice of colleagues (internationally) that we would recruit. I was one of the first to be involved, a co-founder of the department. We sought to create something new, an interdisciplinary department comprising colleagues from all the surrounding disciplines that had genuine interests in Materials. These include Structural Materials (evolved out of what had previously been known as Metallurgy), Electronic Materials (our focus was on Compound Semiconductors at the insistence of Prof. Herbert Kroemer who, as noted earlier, was awarded the Nobel Prize in Physics in 2000, the same year of my award) and Polymers, otherwise known as macromolecules.

In 1986, in the process of building the macromolecular division of our newly formed Materials Department at UC Santa Barbara, we convinced Paul Smith to leave DuPont Central Research and come to UCSB. Whereas I and MacDiarmid and most of the early players in the conducting polymer field were amateurs in the field of polymer science, Paul was a professional. He quickly hammered into my head the importance of making conducting polymers soluble in common solvents and thereby processible from solution. Processing from solution is especially important for creating the uniform thin films that would become the most important format for the many applications that subsequently evolved for semiconducting and metallic polymers.

Paul had an annoying habit of asking me questions for which I had no answer. He wanted to know the intrinsic conductivity, the highest electronic conductivity that could be achieved in metallic polymers. The analogous question — What is the ultimate strength of a polymer fiber? — had been answered by Paul and colleagues at DuPont. With help from Dr. Steven Kivelson, a physicist friend of mine (now a Professor of Physics at Stanford University), we were able to answer the question with surprising results. For well-aligned chains of polyacetylene, the prototype metallic polymer, we found that the maximum conductivity at room temperature should exceed that of copper. Moreover, under those conditions, the strength of such polymer fibers would exceed that of steel. This remarkable prediction was in fact verified by a number of researchers around the world. My colleague Shirakawa provided the data on the high strength; we and a group in Germany demonstrated electrical conductivities approaching that of copper (actually greater than that of copper on a mass basis).

Paul and I began to think about starting a company, to be the first to commercialize products made from conducting polymers.

We were aware of a group from Neste, a petrochemical company in Finland. They had made impressive progress. At a meeting of the American Chemical Society in Dallas, Jan-Erik Osterholm showed large sheets of polythiophene which were truly impressive. Over drinks that evening, we expressed our desire to start up a company in Santa Barbara to do research and development, realizing (and admitting) that any successful commercialization would require the assets of a major chemical company. We invited him to join us in Santa Barbara for a few days to discuss the idea in greater detail. Following these initial discussions, a plan was made. Jan-Erik was enthusiastic and promised to present the idea to the higher management at Neste in Helsinki. Not long after, I received an e-mail indicating that Neste would back us in creating our new start-up company. Much had to be done. After getting advice from Herbert Dwight, the founder and creator of Spectra Physics, we hired a very high-powered law firm from Los Angeles and began the necessary negotiations.

Paul and I struggled over the name. In the end, however, the data provided the name. These high-performance numbers that I alluded to above required highly aligned polymer chains which would have uniaxial properties; the electrical, optical and mechanical properties of such oriented materials are uniaxial. Thus, in a coffee bar in Dulles Airport after a long day of attempting (successfully) to raise money for our university research, Paul suggested "UNIAX" as the company name. It was the right name and was immediately accepted by both of us. We sketched out a logo on a napkin, and the company was born.

The deal was made. Neste would value the start-up at $1,500,000 and would provide us with that amount in cash. During a visit by the Vice-President, we convinced Neste to remodel and purchase a small building for our new company located only a few minutes from campus. I could move from my office to UNIAX in only four minutes.

Our first employee was Dr. Yong Cao. I first met Yong in 1986 at a conference in Beijing during my first trip to China. I had interesting and important new data to report at the conference and was looking forward to showing off these new results to the international group of scientists in attendance. Yong Cao was scheduled on the program to give his lecture before mine. To my dismay (and delight), he had precisely the same results that I was planning to present. I was truly impressed. In 1986, science in China was at a very low level. Nevertheless, Yong Cao had matched my efforts with minimal equipment and in a relatively unsophisticated environment.

We brought Yong Cao to our lab as a post-doctoral researcher within a few months following the conference in China. I did not know just how good he was — but quickly began to find out. At that time, conducting polymers were generally believed to be insoluble. Almost immediately after his arrival, Yong demonstrated that polyaniline could be dissolved and processed from solution. Yong had done his PhD studies in St. Petersburg (then Leningrad) during the years of close friendship between China and the Soviet Union. After returning to China, he was faced with Mao Tse-tung's Cultural Revolution, during which he spent several years working in the fields in the countryside. He then did the equivalent of a second PhD under the direction of none other than Prof. Hideki Shirakawa at the Tokyo Institute of Technology. Somehow, the world is small and our future collaboration was meant to be. Yong's creativity, determination and scientific strength were critical to our scientific progress and to the success of UNIAX. During the 1990s, UNIAX played a leading role in developing the science and technology of conducting polymers.

Yong Cao was the first employee at UNIAX. His first accomplishment at UNIAX was to discover a method to process semiconducting and metallic polymers from solution. The ability to process a metallic polymer from solution was a genuine breakthrough. We built up a research staff that eventually grew to approximately 35 before the company was acquired in 2000.

Yong returned to China in 1999. He is now a leading scientist in China, a member of the Chinese Academy of Sciences, a member of the World Academy of Science, and the Director of a major institute in Guangzhou.

This was our first start-up. We were naïve, and the business plan (when I look back on it now) was a joke. Even the deal with Neste had fine print with critical conditions of which we were not aware despite the advice of our high-priced LA lawyers. A particularly important specific clause that passed by our attention was that we were obligated to pay back the $1,500,000 within seven years or all assets of the company would return to Neste. Missing that condition was indeed very stupid on our part, but as it turned out we were able to meet this obligation and more.

We had to be very careful of our relationship with the university. What was done at UNIAX and funded by UNIAX belonged to UNIAX, but since we were professors at UCSB, we were obligated to report any patentable discovery to the university and obtain a release from the university. What was done at the university belonged to the university. However, we were able to license the university patents into the company.

We therefore licensed our initial patents based upon work carried out at the university from the University of California, and for consistency, agreed to use the method developed by the University of California as a means of putting value on any patents that were developed at UNIAX. This valuation was based on projected sales of products resulting from that patent over a specific period of years. As part of the founding agreement, Neste agreed to take a license from UNIAX (based on our first patent on the solubility of polyaniline) when the appropriate level of development indicated that the technology was ready for large-scale commercialization. The terms of that license included royalty payments and an upfront licensing fee. The magnitude of the licensing fee was to be based on the size of the market, as in the UCSB model.

For the first four years of the UNIAX era, we had two scientists from Neste in residence working with us toward this goal. These two colleagues shared our enthusiasm for the progress at UNIAX. We often discussed the possible applications and the implied value of our technology. There was no disagreement; with each step that we made toward stable, processible metallic polymers, our colleagues from Neste increased in their minds (as did Paul and I) the value of the technology.

Significant improvements in the materials were made at UNIAX. I took these improved polymers back to our university labs where the higher quality enabled us to make important progress on the basic science. Thus, we benefited from UNIAX in two ways. We enjoyed financial gain when UNIAX was acquired in 2000, and we also were able to continue to be the scientific leaders because we had access to the best materials available.

When we were convinced that the technology was ready for scale-up and commercialization, a meeting with the senior management of Neste was arranged to be held in Helsinki. The initial discussions went badly. The two Neste scientists who had worked with us and who knew (and agreed with us in earlier discussions during the four years of technology development) the potential value of the market, presented an analysis which indicated a much lower market size than we had anticipated — implying a significantly reduced upfront payment.

I am not a good negotiator, but fortunately Paul was strong and defiant under these conditions. He looked the senior Vice-President in the eye and told him in no uncertain terms that he had been given incorrect information — that his junior colleagues had lied to him. There were threats of a lawsuit that we were convinced would be upheld in the California courts. In the end, we came to agreement and a term sheet was written and signed. Neste agreed to an upfront payment of $5,000,000. In addition, to reward us for our success and to prevent any subsequent legal actions, Paul and I were each offered a consulting contract that would provide $500,000 over a period of three years. We celebrated that evening; we had passed a significant milestone for the company, and we had each done well in financial terms.

Several important events took place during the first five years of the life of UNIAX. In the summer of 1992, the group at Cambridge University under Richard Friend discovered electroluminescence from thin films of semiconducting polymers, i.e. polymer-based Light Emitting Diodes (LEDs). Neste sold its shares of UNIAX to Phillips (the Netherlands) and to Hoechst (Germany) in 1994. And in 1995, Paul received an offer from the ETH in Zurich that he could not refuse. He sold all his stock back to the company for $100,000, thereby significantly increasing my fraction of the ownership.

The $5 million was used to restart the company in a new direction — to develop polymer LEDs with an initial goal to create displays for mobile phones. I had been CEO (and Chairman) of UNIAX for the first four years, and in each of these years the company showed a profit. The profit did not come from commercial sales, but from contracts with the U.S. government and from additional support from Neste resulting from research proposals that I continually wrote requesting funds for specific projects.

I stepped down from the CEO position and the board hired a seasoned high-tech businessman, Mr. James Long. We moved to a new and larger facility, installed the necessary equipment including a clean room and a processing line for carrying out the multiple steps required for making the LEDs. Our particular "unfair advantage" was that we immediately understood the importance of processing the light-emitting polymers from solution. We filed an important patent on the fabrication of polymer LEDs with each step utilizing the processing of thin films from solution. This patent turned out to be of principal importance to the final valuation of UNIAX.

UNIAX had a strong Board of Directors, including Paul Smith, me and Jim Long. In addition we had a representative from Phillips, Matthew Madeiros, a representative from Hoechst, Rupert Rupp, and as the independent Director, William Cook. Bill Cook was tough and sometimes difficult to deal with, but the Board functioned well.

In 1998, I became disillusioned with the way that Jim Long was running the company. I expressed my concerns privately with each member of the Board. The decision was made. Jim would be fired, and I would again become CEO.

My second term as CEO was far more difficult than the first. I was run over the coals many times for not properly running the business. I was saved by a telephone call from a former student of mine, Curtis Fincher, then a senior member of the research staff at DuPont. Curt invited me to Wilmington to give a lecture. DuPont was in the mood to move up the "food chain" — to supply not only materials as had been their traditional business, but to develop and market high-technology products. The UNIAX polymer-based LEDs for use as displays for cell phones was immediately appealing to them, as shown in the photo below.

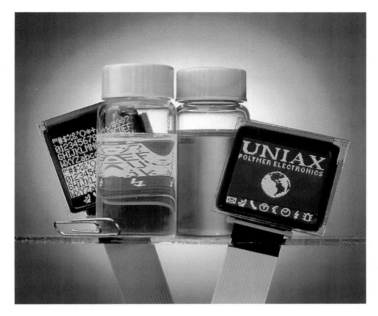

The different colors on the two displays come from two different
luminescent polymers.

One of Jim Long's most important contributions was the hiring
of Boo Nilsson as a Senior Engineer. Boo created the pilot line that
was used to fabricate the displays. On the particular day of the first
visit by senior people from DuPont, we had just finished an impressive
display — 128 by 128 pixels, and they all worked beautifully. We could
program images and messages on this display. The discussions with
DuPont became immediately serious.

I then initiated formal discussions, and we had a Letter of Intent in
August, 1999. I quickly realized, however, that I did not have the experience to handle such a major transaction. Bill Cook had that kind of experience, and he was made CEO and given the responsibility to handle the
acquisition of UNIAX by DuPont.

There were many complications, "ups and downs," and many times
when I thought the acquisition would never come to fruition. The final
papers were finally signed in March 2000, and the transfer of funds
occurred shortly thereafter. I was indeed a Happy Man! At an early stage

in the history of the company, I had given each of my sons and their families stock in the company. Peter received sufficient funds from the DuPont transaction to educate his two sons through Ivy League universities, through Harvard Law School (Brett) and through graduate school in accounting (Jordan). Jordan is now a CPA and working hard in a start-up in Seattle. [David's two girls are approaching their university career in 2015].

A few of my students or post-docs have followed my example and founded successful high-tech start-ups. The most recent is Dr. Anshuman Roy. Anshuman did his undergraduate and graduate studies in India and came to the US for post-doctoral studies. He worked with me for two years. Anshuman is very smart and could have successfully pursued an academic career. However, he wanted to start a company. He even had a name for his company that he carried with him from his childhood, Rhombus. He succeeded in starting Rhombus based on the invention of a high sensitivity neutron detector capable of detecting fissionable material. Fissionable material can be used to make nuclear bombs; fissionable material emits neutrons. He developed his invention into a high-value product. His device will be employed in ports and transportation hubs all over the world to prevent terrorists from importing fissionable material. As a result, people all over the world will be safer.

Doing a start-up is very exciting, but very stressful. There were times when I had sufficient funds in the treasury to make the payroll for only the next few weeks. One must have a technology that is truly new (like our conducting polymers) or something that is both innovative and has a clear established need. One must learn to work with the Board of Directors whose responsibility is to maximize the value to the shareholders. Now I was a successful entrepreneur. I thought that I had made all the possible mistakes and was therefore ready and even eager to start up a second company, I would find that I still had a great deal to learn.

My colleagues at UNIAX, in particular Dr. Gang Yu, fabricated some very special displays for me to take with me to the Nobel festivities.

The image of Alfred Nobel on a UNIAX display was shown during the Nobel Award Ceremony.

I still have this organic LED image of Nobel, and it continues to work even after 15 years have gone by.

My only possible regret is that the successful acquisition of UNIAX took place in March, whereas I was informed of the Nobel Prize in October. One wonders whether the valuation would have been higher if these events had occurred in the opposite time order! I am certain that the UNIAX success had a major influence on the award of the Nobel Prize. We had not only created a new field of science, but we had demonstrated that this field would generate the new era of "plastic electronics."

17 The Impact of Theater in our Lives

Niels Bohr was one of the greatest physicists of the 20th century. His theory of the hydrogen atom, published in 1913, was a major step toward the development of quantum mechanics. Bohr was a Dane and lived in Copenhagen with his wife, Margrethe. He was known as being particularly careful; it is said that he typically wrote 17 drafts of every paper, each of which Margrethe typed in full, before he was satisfied that the article was ready to be sent for publication. He also loved to discuss physics with his colleagues, and spent many hours "decanting" his results again and again to assure clarity. Bohr was awarded the Nobel Prize in Physics in 1922. The Danish Government founded an Institute of Theoretical Physics to honor Niels Bohr name; The Bohr Institute in Copenhagen.

Among the long list of physicists who came to work with Bohr was Werner Heisenberg, a young German. Heisenberg was a genius; he finished his PhD studies in his early 20s. He came to work as Bohr's assistant in 1924 and remained at the Bohr's Institute for three years from 1924 to 1927, during which time he and Bohr published many important papers and developed the Copenhagen interpretation of quantum mechanics.

Heisenberg received the Nobel Prize in Physics in 1932 for his discovery of the uncertainty principle; a quantitative statement that one cannot simultaneously know both the position and the momentum of a particle, as expressed by his famous equation, $\Delta \times \Delta p \leq \hbar/2$, where $\hbar$ is Planck's constant (6.6×10^{-34} m^2 kg/s) and Δx and Δp are the uncertainty in the position and the momentum of the particle, respectively.

During their years together, the Bohr–Heisenberg relationship became almost like that of a father and son. In 1927, Heisenberg received a Chair at Leipzig University and returned to Germany. They continued to interact until the outbreak of the Second World War, when Denmark was quickly overrun and occupied by Germany.

In 1941, Heisenberg returned to Copenhagen. The reason for his trip remains uncertain, and what occurred during his three-day visit is a matter of genuine interest and importance to the history of physics.

Nuclear fission had been discovered in 1937. The concept of a nuclear bomb was proposed shortly thereafter by Leo Szilard. Szilard and Enrico Fermi visited Einstein in 1939 with the express purpose of warning him of the possibility of a nuclear weapon, and to inform him of their suspicions that the Germans were already at work developing this new method of mass destruction. Szilard and Femi made a special request: They asked and convinced Einstein to write a letter to President Roosevelt informing him of this impending danger. The Einstein letter to Roosevelt in 1939 was sufficiently influential that it subsequently resulted in the creation of the Manhattan Project and the U.S. development of nuclear weapons.

It seems quite clear, even obvious, that Heisenberg's return to Copenhagen in 1941 had something to do with the Bomb. His visit, therefore, was a dramatic event. Michael Frayn saw the drama in that visit, and he wrote his remarkable play *Copenhagen* about that visit. *Copenhagen* was produced in London and later on Broadway. The Broadway production received three Tony Awards: Best Play (Michael Frayn, playwright), Best Direction (Michael Blakemore) and Best Performance by a featured actor (Blair Brown).

Frayn is a playwright — he writes fiction; he is not a scientist. Nevertheless, he researched the Copenhagen visit in sufficient detail, that he got the physics right. I know that is true, because I had the opportunity to play the role of Niels Bohr in several productions of staged readings of *Copenhagen*. My colleague and Nobel Laureate David Gross took the role of Heisenberg. Bohr's wife, Margrethe, was played by Stephanie

Zimbalist in the first production, which took place in Santa Barbara. In two subsequent productions, one in Brussels on the occasion of the centennial celebration of the famous Solvay Conference of 1912 and the second as part of the Nobel festivities in December 2013 (the centennial of Bohr's theory of the hydrogen atom), the role of Margrethe was played by Fiona Shaw, a well-known and very accomplished British actress with many credits to her name.

Our *Copenhagen* saga began when Nancy Kawalek, a friend of ours and a member of the Board of Directors of the Ensemble Theatre Company in Santa Barbara, suggested to the Board that since we had five Nobel Laureates on the UC Santa Barbara faculty, there was an opportunity to create a fund-raising event with a staged reading of *Copenhagen* with Nobel Laureates playing the roles of Bohr and Heisenberg. She came to Ruth with the idea. Ruth was enthusiastic; she spoke to me about this opportunity, and I immediately agreed to participate and to take on the role of Niels Bohr. Ruth then contacted David Gross (Nobel Prize in Physics in 2008 for his discovery of the "strong force" that holds nucleons together). David was equally enthusiastic, but insisted that he wanted the part of Werner Heisenberg. Sometimes things just work out right. Nancy is a theater person, an actor and a director; she held a faculty position at UC Santa Barbara. Soon thereafter, Nancy contacted Stephanie Zimbalist, who agreed to work with us and take the role of Margrethe, Bohr's wife.

A date was set, and David and I began rehearsing under Nancy's direction, sometimes with Stephanie, but more often with a stand-in because we were truly amateurs and had a great deal to learn about acting even in the setting of a staged reading. The staged reading would take place at Hahn Hall, a beautiful 300-seat theatre at the Music Academy of the West, in Santa Barbara, on Sunday March 7, 2010.

Ruth considered it to be her responsibility to fill the house. The production was announced and locally advertised. But Ruth personally sold many of the tickets; she went to "angel" donors who were convinced by her passion and gave generously. A few days before the performance, there were only three open seats remaining; she sold these for $500 each. Thus, we were faced with a full house. Nearly $80,000 was donated to the Ensemble Theatre from the proceeds.

Many of my physics colleagues were dubious about our ability to pull this off. After the performance, I received genuine and hearty congratulations (!) from these same doubters and many more from others who had been in the audience. We have now done the staged reading of *Copenhagen* three times, each time better and each time with more sophistication and more realistic acting. Perhaps this is becoming a second career for David and for me.

Perfect days! I listed a number of perfect days in Chapter 1. The opportunity to play the role of Niels Bohr in *Copenhagen* has now generated several more perfect days. Moreover, by doing the Bohr role and listening carefully to the words spoken by Heisenberg, I learned some new physics that I would use to solve a long-standing problem — more about that in the following chapter.

Michael Frayn came to the performance in Brussels and again to the performance in Sweden. He told us afterward that he particularly enjoyed having physicists in the roles of Bohr and Heisenberg. One could tell the difference, he said, when the actors really understood the physics. He also said that if he were ever to revive the play in London or on Broadway, he would want real scientists as the actors.

Our involvement in the theater did not begin nor end with *Copenhagen*. Ruth and I had long been supporters of the Ensemble Theatre Company in Santa Barbara. We became season ticket holders for the first season following our arrival in Santa Barbara. Although the theater was not as professional then as it is today, we enjoyed many outstanding performances. I remember that I could hardly breathe after the last scene in *Death of a Salesman*. I commented earlier about the impact on me of Terrence McNally's play *Master Class* — one must never lose one's nerve. Our involvement increased as we became serious donors, and both Ruth and I served for a time on the Board of Directors.

The venue for the Ensemble Theatre Company (ETC) was hopelessly unsatisfactory. There were no dressing rooms, the restrooms were inadequate and the seats were seriously uncomfortable. The dream was to create a new theater as the home base for the ETC. This dream began to take on serious reality after Jonathan Fox was convinced to come to Santa Barbara and serve as the Artistic Director of ETC. The ETC immediately became professional, with actors and directors brought in by Jonathan from New York and Los Angeles. The need for a new theater became even more evident.

The ETC is now in a new theater in the heart of the theater district of Santa Barbara, the "New Vic." Nearly $15 million was raised from the Santa Barbara community to create this new and beautiful 300-seat theater, with Derek Westin taking the lead in raising the necessary funds. There are excellent and functional dressing rooms, there are plenty of restrooms for the audience, the seats are very comfortable, and the "rake" of the theater is sufficiently steep that one can see clearly and enjoy the performance from any seat in the house.

Ruth and I again participated in bringing this about. We donated generously to the building fund, and Ruth designed and supervised many successful fund-raising events.

But our involvement in the theater does not stop with the success of the ETC at the "New Vic." We have always been Broadway fans, and we made a point of going to the theater whenever we had the opportunity to visit New York or London. Nevertheless, we did not envision having the pleasure of involvement in the production of several Broadway shows. Our friend Walt Grossman and his colleague Robyn Goodman have enabled us to invest and participate in the production of four Broadway plays.

The first was a revival of *Barefoot in the Park*. We attended the opening night performance, laughed loudly and truly enjoyed this wonderful comedy. After the show, we went to the traditional party during which one waits (as in the old movies) to learn of the judgement of the critics. Unfortunately, the *New York Times* critic Ben Brantley's review was not just bad, it was vitriolic! A bad review from the *Times* is typically sufficient to close the show. *Barefoot in the Park* survived, however, for two months, sufficient time for us to recoup more than half of our investment.

Our next Broadway theatre adventure was an original musical called *In the Heights*. The music was original and good, part hip-hop, part jazz and part just good, and the dancing was fantastic. *In the Heights* received 13 Tony nominations and was the winner of seven Tonys. We were elated and had a great time at the after-theater party. The show had a two-year run on Broadway and subsequently went on tour, so we continued to receive checks for approximately three years.

We were involved in the revival of *West Side Story*, the wonderful musical based on the Romeo and Juliet story with music written by Leonard Bernstein. *West Side Story* was again a success, with a two-year run on Broadway followed by a lengthy road tour.

The famous Rodgers and Hammerstein duo had created a musical that had never been produced on Broadway, *Cinderella*. After some rewriting and some modernization of the book, *Cinderella* opened on March 2, 2013. The final performance was on January 3, 2015, again nearly a two-year run. *Cinderella* provided a wonderful evening for young and old.

We continue to enjoy going to the theater at any opportunity. When I experience a really good theatrical experience, it sends a shiver up my spine. We continue to support the ETC in Santa Barbara as season ticket subscribers and through philanthropic contributions such as sponsoring a specific play. Ruth actively "recruits" new subscribers by purchasing extra tickets for each performance and inviting a new couple to join us for the opening night festivities.

Thus, the theater has played a significant role in our lives; it has enriched the quality of our lives far beyond what I could have imagined after seeing our first Broadway production in 1957.

18 Low-Cost "Plastic" Solar Cells – A Dream Becoming Reality

Solar cells convert sunlight to electricity. The energy from the sunlight that falls on the Earth in one hour is sufficient to satisfy the energy needs of our planet for one year. The power of the sun is a truly renewable source of energy. Thus, the discovery of the silicon solar cell at the Bell Laboratories in 1954 provided a pathway to renewable — and non-polluting — energy.

Solar cells had their first major impact in the space program. The simple batteries that were used in the earliest satellites were quickly drained. Solar cells, however, can be run continuously in space for years. Thus, the beginning of public consciousness of solar cells came from the success of the U.S. space program.

Only much later, when concerns of the long-term availability of fossil fuels became a subject of debate, were solar cells considered seriously as a source of electricity. The dire predictions of global warming as a result of the "greenhouse" effect intensified the quest for renewable, non-polluting sources of energy. Today, solar cells provide a very small fraction of the Earth's energy needs, and most solar cells in use today are made from silicon as the primary material.

Three steps are required for generation of electrical power through the absorption of incident sunlight:

(i) Absorption of photons by the photoactive material;
(ii) Photoinduced charge separation and the generation of mobile charge carriers;

■ 213

(iii) Collection of electrons at one electrode and holes (electron states that are unfilled and therefore respond as if they were positively charged) at the opposite electrode.

These three steps are accomplished in all photovoltaic cells whatever materials are used as the photoactive materials in the solar cell.

The concept of "plastic" solar cells is appealing for many reasons. They are lightweight, flexible and rugged. They can be manufactured by low-cost printing technology, and they offer a variety of novel applications, including semitransparent widows that both generate electricity and reduce the need for air-conditioning by reducing the amount of warming sunlight that enters the building through the windows; lightweight plastic solar cells can be mounted on simple mechanical structures that would not support the weight of silicon solar cells — and many more. Some of these opportunities are illustrated in the figure below:

Bulk Heterojunction Solar Cells

Our research in the general area of "plastic electronics" led us naturally to consider the possibility of thin film plastic solar cells made from

Plastic Solar Cells

Nano-OPV

Lightweight, flexible and rugged !

Example of Plastic Bulk Heterojunction Solar Cells and their unique opportunities for novel.

semiconducting polymers. The initial discovery of ultrafast electron transfer between semiconducting polymers and fullerenes occurred in late 1992. It was a discovery based purely on curiosity. At that time, we had been working on the optical properties of semiconducting polymers for many years. Then, the remarkable "buckminsterfullerenes" — molecules with the shape of a soccer ball comprising 60 carbon atoms — were discovered by Richard Smalley, Harold Kroto and Robert Curl; they were awarded the Nobel Prize in 1996 for their discovery.

Prof. Serdar Sariciftci was a post-doctoral researcher in my group in 1992. During a random discussion in my office, we speculated on what would happen if we mixed these two novel materials (soluble semiconducting polymers and soluble fullerene derivatives). We made several speculative guesses, but decided to do some initial experiments even though the idea was not yet well formed in our minds. We obtained the now famous soluble fullerene derivative, nicknamed PCBM, from my friend and colleague Fred Wudl, and the story began to unfold.

That the luminescence of the polymer was heavily quenched by the addition of fullerenes suggested that electron transfer from the polymer to the fullerene must occur on a time scale significantly faster than the decay time of the photoluminescence, i.e. at least in the picosecond time regime. A picosecond is 10^{-12} seconds — a billionth of a billionth of a second! Thus, we decided to measure the electron transfer time directly using ultrafast pulsed laser techniques. The result of these initial ultrafast experiments demonstrated that the photoinduced electron transfer occurred in < 100 fs i.e. < 000000000000001 seconds! The entire field of plastic bulk heterojunction solar cells was created as a result of this demonstration of ultrafast charge transfer. Since the electron transfer rate was orders of magnitude faster than any competing process, we inferred that the efficiency of photoinduced charge generation must be nearly 100%, implying the possibility of high-efficiency solar cells. The first bulk heterojunction solar cells (using PCBM as the acceptor) were fabricated and the results reported in 1993.

A bulk heterojunction (BHJ) material is a solid state mixture of two components (a donor of electrons and an acceptor of electrons) with nano-structured morphology formed by spontaneous phase separation: The donor and acceptor components self-assemble to form bicontinuous interpenetrating networks. If the separated electrons and holes recombine and return the material to its lowest energy state (the so-called ground state), mobile charge carriers are lost and cannot contribute to the generation of electricity from absorption of the incident sunlight. Because the carrier recombination lengths are typically approximately 10 nm in these disordered materials, the length scale for this self-assembly must be of order 10–20 nm (10–20×10^{-9} meters).

The formation of two interpenetrating networks requires that the component materials phase separate, that the interfacial energy favors high surface area and that each of the two components is fully connected with uninterrupted pathways to the electrodes. All of this happens spontaneously by <u>self-assembly</u>. Truly remarkable and not previously anticipated!

The bulk heterojunction (BHJ) plastic solar cell is an example of what is known more broadly today as "nanoscience," that is, science carried out on structures that are much larger than atoms or molecules; structures made up of atoms and molecules where the size scale is nanometers (1 nanometer is $0.000,000,000^{th}$ of a meter). Nanoscience became a buzzword in the scientific community in the mid-to-late 1990s. Because of our early work on the plastic solar cells, we were doing research on nanoscience well before it had become established in the scientific community as an important and challenging opportunity for research.

Our original idea is sketched as shown in the figure on the next page.

A BHJ material is highly disordered. The components are mixed in a common solvent and then simply cast onto a suitable substrate and allowed to dry. There is typically only relatively minor crystallinity embedded within relatively large regions of amorphous material. In such disordered structures, the electron wave functions are known to be

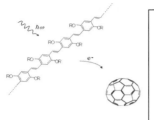

One should imagine the fullerene acceptors as comprising the black network and the donor polymer (or small molecule) as comprising the white material.

Charge transfer plus the formation of bicontinuous interpenetrating and fully connected networks with nano-structured phase separation on the 10–20 nm length scale are required features of the Bulk Heterojunction solar cell.

localized, implying the transport of photo-induced charge carrier to the electrodes, as required in order to deliver power to an external circuit, is hindered by the disorder.

The interface between two different semiconductors is known as a "heterojunction." Because of the different electronic structures of the two semiconductors on either side of the heterojunction, charge separation occurs at the heterojunction boundary. As shown schematically in the concept figure above, there are heterojunctions throughout the bulk of the material. Hence, I chose the name *"bulk heterojunction"* material. As first stated by Tom Friedman: "If you name it, you own it!" So, I conclude that I own the concept of the BHJ solar cell.

The internal electric field created by the charge separation serves to drive electrons toward one electrode and holes toward the opposite electrode. The ultra-fast charge transfer is the key issue. Nevertheless, the mechanism for the ultrafast charge transfer remained a mystery for over 20 years. Only recently has the mechanism become clear to me, although still controversial within the broader scientific community working on BHJ solar cells. Once again — Never Lose Your Nerve.

The ultrafast charge transfer is the result of the fundamental quantum uncertainty as described by Heisenberg's uncertainty principle. This idea was first suggested to me by my post-doctoral research colleague Dr. Loren Kaake. I quickly responded positively because I had been thinking deeply about the meaning of the uncertainty principle as a result of my performances of Michael Frayn's play *Copenhagen*.

At one point in the play, Heisenberg says: "Everyone understands uncertainty. Or thinks he does." Although I had been introduced to the uncertainty principle in my first course in quantum mechanics, I had never thought sufficiently deeply about it to really understand. In the play, Bohr returns to Copenhagen after a week of skiing in Norway. He says, in anger, to Heisenberg:

> By the time I get back from Norway I find that you've done a draft of your uncertainty paper and you've already sent it for publication. My dear Heisenberg, it's not open behavior to rush a first draft into print before we've discussed it together! It's not the way we work!

Heisenberg describes the origin of the uncertainty principle as arising from the measurement process: If you want to find an electron, you try to find it by scattering a photon off the electron. That collision transfers momentum to the electron so that after finding it, one does not know its momentum. That is the way I had first learned the physics of the uncertainty principle.

Bohr, however, goes on to say that the paper contains a fundamental error — that in order to understand the uncertainty, one must use the Schrodinger's wave formulation of quantum mechanics. Heisenberg responds: "I know, Complementarity. I know — I put it in a post-script to my paper." Bohr ends the discussion by saying that everyone remembers the paper, but no one remembers the post-script.

So, naturally, I went to the original publication written of course in German, and I read the post-script. Only then did I begin to understand what Bohr meant when he said that the paper contained a fundamental error. Remember — Michael Frayn's play is fiction, but he researched the history in sufficient depth that he got the physics about right.

I was prepared. When Loren made the comment about the uncertainty principle as a mechanism for the delocalization that would enable ultrafast charge transfer, I immediately understood and went on to formulate the idea with greater clarity in a draft of a manuscript. We wrote many drafts of that manuscript together, each with more insight. I am satisfied that we finally got the mechanism for ultrafast charge transfer both right and explained with clarity.

We estimated the length scale of the wave function describing the probability amplitude for finding a photoexcitation at a particular point in space in a variety of ways using position-momentum uncertainty, and concluded that the photoexcitation process generates a delocalized wave function that describes the probability of finding the photoexcitation at any specific site. That delocalized wave function is a superposition of the eigenfunctions of the Schrodinger equation that describes the nano-structured organic photovoltaic blend. Thus, there is an immediate probability amplitude for finding a photoexcitation near a heterojunction boundary, thereby enabling ultrafast charge transfer. This conjecture was demonstrated to be true by carrying out transient absorption measurements on a variety of organic bulk heterojunction nano-structured materials to investigate the photogenerated charge transfer dynamics. Startling generality emerged which implied that the lifetime of the delocalized excited state wave function was produced at ultrafast time scales is sufficiently long that it plays an important role during the charge and energy transfer processes critical to ultrafast charge photogeneration.

The data imply that a delocalized state of sufficient lifetime exists in BHJ materials such that it can participate in electron transfer reactions and the charge separation process. As a result, prior to collapse into the localized excitations expected for a disordered nano-structured material, electron transfer can occur from phase separated domains of electron donating materials (polymers or small molecules) to domains comprised of electron accepting materials (typically substituted fullerenes) in the femtosecond time regime.

A comprehensive analysis of the optimized power conversion efficiencies has been carried out under a set of reasonable assumptions. Based on this analysis, we should expect to be able to achieve power conversion efficiencies (PCE; power out/incident solar power absorbed) of 20% or possibly even greater. To achieve this high power conversion efficiency will require the demonstration of external quantum efficiency (EQE) = 90% and a fill factor of at least 0.7. Fill factors > 0.7 are commonly observed in recent publications. EQE = 80% has been demonstrated. The use of higher-quality transparent electrodes should be able to increase the EQE to 90%.

The achievement of plastic BHJ solar cells with 20% power conversion efficiencies will be a major accomplishment with important consequences to energy technology.

a. Low $ cost of manufacturing
b. Low energy cost of manufacturing (no high-temperature steps are required).
c. Low carbon footprint manufacturing
d. Flexible, lightweight and robust solar panels

The implications of roll-to-roll slot coating or printing technology on flexible plastic film cannot be overestimated. These "printing" techniques are low cost and capable of producing huge areas (just think of newspaper printing!). Consequently, plastic solar cells can be manufactured in areas considerably larger than the silicon panels or related solar cell technologies available today.

Although I express optimism of this coming to fruition, the power conversion efficiency of BHJ solar cells is currently too small for large-scale implementation. Although the progress is impressive — from 1% in 1993 to nearly 12% in 2014 — we are not yet ready for large-scale commercialization. In our labs, we have demonstrated 11.3% PCE (July 2014); the highest value that has been reported is 11.7% PCE. The scientific community must increase the PCE to 15–20% in order to enable large-scale introduction into the market place. Will this actually happen? I do not know, for that is the inherent uncertainty in science.

The vision, however, is clear: I foresee plastic solar cells on roof-tops, in windows (semi-transparent) and on the facades of buildings all over the world.

Konarka Technologies, Inc.

Sukant Tripathy was a long-time friend, a scientist whom I greatly respected. He typically did not follow the paths created by others; he went in orthogonal directions that often proved both creative and productive.

In the Fall of 1999, Sukant approached me with the idea of starting a company with the goal of creating a technology for low-cost solar cells. Although I was immediately interested, I was in the midst of the acquisition of UNIAX by DuPont, and simply did not have the time or energy to take on such a new endeavor. I told him to postpone this until after my dealings with DuPont had come to fruition.

Then, in mid-October of 2000, I received the life-changing telephone call from the Nobel Foundation. I received many, many notes of congratulations including a note from Sukant. I asked him to be patient. I would be ready to join him in this interesting endeavor after the beginning of the new year.

In December of 2000, I went to Sweden to receive the Nobel Prize. In December of 2000, Sukant went to Hawaii to attend and participate in the Pacific Rim Conference of the American Physical Society. Unknown to me, Sukant drowned while swimming at the black sand beach on Oahu.

When I returned to Santa Barbara after the Nobel festivities and after a week of skiing with my family, I found a strange e-mail in my "inbox". I was invited to give a lecture at a memorial conference in honor of Sukant Tripathy. I did not know what had happened, but quickly learned the sad details. I agreed to come to Lowell, MA, to give the lecture. The furthest thing from my mind was the unborn plan that we had discussed concerning the start-up of a new company to create the technology for low-cost plastic solar cells. Without the founder, there could be no such company.

After my lecture, I was invited to a lunch meeting with the Chancellor. At this meeting, I met Howard Berke. Howard was a serial entrepreneur and had agreed to serve as an advisor for the University of Massachusetts at Lowell. He had many discussions with Sukant and was convinced of the opportunity. I repeated my conclusion: "Without the founder, there can be no company." But Howard was determined. He had already had initial discussions with venture capital companies, and he had identified an initial team, some of whom had worked as students or post-doctoral researchers for Tripathy. He would not let go and eventually convinced me to be a co-founder, Chief-Scientist and to serve as a member of the Board of Directors. Howard served as Chairman. The company was named after a Temple of the Sun God near the village in India where Sukant was born: Konarka Technologies, Inc.

The company grew with specific emphasis on bringing aboard engineers with experience in roll-to-roll coating. Interesting plastic solar cell samples began to appear, but with very low efficiencies. The scientific staff was enlarged with the goal of improving the power conversion efficiency, but progress was slow. Konarka was stuck at efficiencies just below 3%, too low for any commercial application. Meanwhile, scientists worldwide began to work on this interesting technology. The efficiency gradually improved, but not fast enough.

The death of Konarka was caused by a risky decision. The Polaroid Corporation had recently gone into bankruptcy. Polaroid had built a large manufacturing facility for roll-to-roll coating of photographic film at a cost of approximately $250 million. Since the Polaroid films comprised many layers, it seemed that this manufacturing facility would be perfect for producing plastic solar cells. After some initial test runs, we were convinced that the former Polaroid plant would work for Konarka. Konarka acquired the facility for less than $5 million, and part of that was subsidized by the state of Massachusetts as a means of creating new jobs.

The Konarka manufacturing facility was impressive, and it actually worked. The yield was as high as 75%, quite remarkable for a new technology. The Konarka staff worked hard to improve the efficiency and actually demonstrated nearly 9% in the lab with dimensions of 1 cm^2.

However, the cost of running the facility was simply far too high without useful (and saleable) solar cell products coming out the end of the line. The electrical and heating costs were high. The cost of the semiconducting polymers was high. Thus, despite Howard's miraculous ability to raise more money, Konarka went into bankruptcy near the end of 2011.

To me, that bankruptcy was like a death. I had a great deal of trouble dealing with the reality of this loss. But, the vigor in the research community, the progress in my own lab and the evolution of strong interest in the products that Konarka had demonstrated revived my

Parking structure with plastic BHJ solar panels fabricated by Konarka, Inc.

own interest. The field of plastic BHJ solar cells goes on with deeper understanding of the basic issues and with correspondingly increased power conversion efficiencies. I am convinced that we will see these plastic solar cells in a variety of applications in the years to come. Konarka had a great vision, Konarka created some remarkable demonstrations (such as the one above), but Konarka started too soon. The field of BHJ solar cells was in its infancy in 2001, and Konarka did not have the resources to see it through to maturity.

19 The Serial Entrepreneur

September 11, 2001 — The Birth of CBRITE Inc.

On September 11, 2001, I was in New York with my friend and colleague Nick Colaneri. We were there for a business meeting with the goal of starting a new company. Shortly after the award of the Nobel Prize, I had been contacted by a group of wealthy potential investors, all of whom were alumni of the University of Pennsylvania. Their intention was to convince me to start a new company based upon our Nobel discoveries. I explained that we had already done that, in part, and gave them a short history of UNIAX. I went on to explain, however, that UNIAX had focused on the front plane of a display (the light-emitting polymer layer) whereas the real value was in the backplane.

If you look closely at an LCD or an OLED display, you will see that the picture is not continuous, but divided into tiny "pixels." In each pixel, behind the light-emitting front plane on which you see the image, there is a tiny circuit (comprised of a few transistors and capacitors) that turns the pixel on and off. The circuits within a pixel must be as small as possible, for they take up "real estate" that would otherwise contribute to the light emission. These tiny pixel-controlling circuits are connected in columns and rows into what is known as the "backplane" of the display. The backplane is the most important element of a display; the backplane creates the images that one sees. Nearly half the cost of the display is in the backplane.

Backplanes for liquid crystal display (LCD) are currently made from amorphous silicon. The new organic light-emitting diode (OLED) displays that we at UNIAX and other researchers and companies world-wide had developed, require significantly higher current than can be provided by amorphous silicon — hence an opportunity. There are other backplane opportunities as well, for example e-books, which are so commonly in use today. Backplanes for e-books and related displays do not require higher performance then available from amorphous silicon, but they are relatively expensive.

Boo Nilsson, Nick Colaneri, Gang Yu and I had a vague plan for a second company that would provide low-cost backplanes for electronic shelf labels (for example, in supermarkets) e-book type displays, and subsequently move up the chain to solve the problem of the high performance required for OLED displays. Nick and I flew to New York on 10 September for a meeting that was to discuss the necessary aspects that must be decided and agreed upon at the moment of the start-up of a new high-tech company. Our 10:00 am meeting was to be on the 16th floor of a building on 52nd Street near Madison Avenue. We stayed the night at the W Hotel on the West Side and walked leisurely across town on a beautiful morning under sunshine and blue skies, stopped for breakfast, and arrived at our destination a few minutes before 10 am. The man at desk whose job was to check out and then announce all visitors was visibly disturbed. While he was taking care of us, he was on the phone, and we heard some comment about the bridges being closed, but without any context, we thought nothing of it.

We took the elevator to the 16th floor. A young woman got onto the elevator on the second floor. She was out of breath, visibly disturbed and said to us that "she had friends in the Pentagon too," and got off at the fifth floor. Nick and I looked each other and said at the same time: "What the hell was that about?" We found out when we arrived at the meeting room and were greeted with the statement: "We are at war. The World Trade Center has been bombed." After some discussion of what had happened based on the early news, the group of approximately 10 people sat down with the intent of having our meeting despite the disturbing news of what was happening downtown. Strangely, there were no television sets in the office, but every few minutes one of the secretaries would come in with evermore disturbing news. The buildings were not actually bombed — full-size commercial airliners loaded with jet fuel had crashed into them. Fires had started and evacuation was already underway. However, it seemed that the upper floors were cut-off by the fires with no means of evacuation. Minutes later, we were told that the first tower had collapsed and then again a few minutes later that the second tower had collapsed.

That was the end of our meeting.

I tried many times without success to reach Ruth by telephone. She knew we were in New York for a business meeting and could naturally have thought that the meeting was in the World Trade Center. I did not actually talk to her until several hours later. We were able to reach some-one in Santa Barbara and asked that they call our wives and tell them we were on 52nd Street, miles away uptown, and that we were safe.

There seemed to be no way to get home. The airports were closed and all flights cancelled. All the car rental services in the city had already leased out their full stock of cars. I called a former student of mine in White Plains; miraculously he found and reserved a Hertz car for us at the White Plains Airport. After staying overnight in an upscale apart-ment on Central Park West that belonged to one of our potential inves-tors, we got up at five in the morning, walked down the center of 5th Avenue to Grand Central Station and took the first train north to White Plains. We were able to get the car, but with some difficulty because the Hertz dealer was on the grounds of the airport and the airport was formally closed.

We crossed the Tappan Zee Bridge at 9 am and drove to Daven-port, Iowa, on September 12, where we had a quick dinner and slept for a few hours in a motel. The next stop was in mid-Colorado, near Vail. We planned to stop the third night in Las Vegas, but after arriving in Las Vegas at about 4:30 pm, we decided to go all the way to Santa Barbara. During the entire trip, we were listening to the somewhat confusing news on the radio. Only when I got home did the full impact of the collapse of the World Trade Center and its consequences become real to me.

That was the non-birth of CBRITE. But we persisted and CBRITE, was finally born in 2004, founded by Boo Nilsson, Gang Yu and me with precious help from David Elliman, Walt Grossman and Neal Milch. Using their connections, we sought advice from a venture capital group about a recommendation for a CEO for the new company. The recommended CEO was hired, but turned out to be a disaster. He spent too much money without significant progress, and he was a terrible manager of the CBRITE employees. He was fired by the Board of Directors.

Boo Nilsson had been General Manager at Dupont Displays (after the sale of UNIAX). He was an excellent manager, but I did not foresee his strength as CEO. Similarly, Gang Yu was an outstanding scientist at UNIAX and at DuPont Displays, but I had some doubts that he could rise to the responsibilities of CTO. I was wrong in both cases. Boo has turned out to be an outstanding CEO; he is an excellent manager and a tough negotiator. Boo has developed relationships with some of the largest display makers in Asia. Gang has worked closely with these large display makers with a goal of transferring the CBRITE backplane metal-oxide semiconductor technology while simultaneously managing the technology innovation group in Santa Barbara and filing a large number of patents in the general area of metal-oxide semiconductors for use in backplanes for displays. We were recently told by an executive from one of the large Asian display makers that any company interested in using metal-oxide semiconductors for back planes would have to come first to CBRITE Inc. The CBRITE metal-oxide backplane technology is not only the key to scalable OLED displays, it will be needed for the next generation of LCD displays.

I find it remarkable, but somewhat comforting, that seven scientists and engineers in Santa Barbara (the technology development group under Gang Yu) could create a technology superior to any developed by the multi-billion dollar display companies in Asia — but that is the nature of innovation. Now, after a "rocky start," CBRITE Inc. is on the track to success. I continue to serve as Chairman and to be amazed by the progress in the CBRITE technology.

Cynvenio and CytomX

Science and engineering at UC Santa Barbara is interdisciplinary. As co-founders, I and my colleagues purposely created this interdisciplinary character beginning with the formation of the Materials Department which is comprised of faculty from physics, chemistry, electrical engineering, chemical engineering, mechanical engineering and biology. Many of these individuals hold joint appointments. In the Materials Department, each graduate student is required to have a Principal Advisor and a Secondary Advisor, the second often from a different home department.

The interdisciplinary atmosphere has spread at UCSB through the creation of the Materials Research Laboratory (sponsored by the National Science Foundation), the Institute for Collaborative Biotechnology (sponsored by the U.S. Army Research Office) and the California Nanoscience Institute (funded initially by the State of California). UCSB is recognized internationally for this interdisciplinary approach to science.

When I arrived at UCSB in 1982, I set up my laboratory with the goal of being independent, i.e. having the instrumentation to acquire all the different data that I would need. Today, my students and post-doctoral researchers roam freely. They collaborate with my colleagues and not infrequently publish papers without my involvement (and without listing me as a co-author). Often, my students or post-doctoral researchers will bring me data they have acquired with instrumentation available in the laboratory of a colleague or from a central facility in one of the major interdisciplinary Institutes (the MRL, the ICB, or the CNSI).

Because these are sponsored programs, there is intense competition for being awarded such a block grant. Consequently, there are periodic reviews where faculty summarize their recent progress. It was at one of these reviews that I heard presentations from Professor Tom Soh (Mechanical Engineering) and Professor Patrick Dougherty (Chemical Engineering). Tom gave a fascinating talk on cell separation (and protein separation) using microfluidic technology. Patrick summarized his work on the creation of a large library of proteins that could be sorted by the use of antibodies synthesized with a specific receptor that would bind to the desired protein. They were collaborating in a project with a specific goal of interest to the Army Research Office.

I was impressed and approached them after the session. I invited them to come to my office for a discussion so that I could learn what they were doing in greater depth. Although I was acquainted with both, I had not collaborated with either of them, nor was I a close colleague. Nevertheless, the meeting went sufficiently well that — with considerable enthusiasm — I suggested that we start a biotech company. Thus, CytomX Inc. was born.

I had been approached earlier by a friend in Santa Barbara, Harry Gelles, who said that if I ever started up a biotech company, he would like to be involved and would help us find the necessary capital. We went to Harry's office for an initial meeting and to my surprise and delight found Fred Gluck there with an apparent interest in what we were doing. Fred Gluck is a major figure in the business world. He was Managing Director of McKinsey, the consulting company that serves the largest companies in the world. He had "retired" and moved to Santa Barbara. Fred is recognized for his role in strategic thinking in Walter Kiechel's book, *The Lords of Strategy*. When at a dinner party, I asked Fred how Gerstner had been able to turn around the behemoth known as IBM, he responded that Gerstner's first move had been to contact Fred Gluck.

Of the many things that Fred had done in the business world, he had never done a high-tech start-up. Thus, he was immediately interested, and my credentials gave him some confidence that we were not just bozos with a half-cocked idea. Before the meeting was over, Fred agreed to participate. He had, however, two conditions: He insisted on being Chairman of CytomX, and he insisted on being CEO. I was delighted! To have someone of his stature and proven quality at the helm was a major asset. Thus, Fred Gluck, Alan Heeger, Harry Gelles, Tom Soh and Patrick Dougherty co-founded CytomX LLC.

Given Fred's background, it is not surprising that while beginning to create a laboratory and hire the necessary initial research staff, we spent many, many hours discussing the strategy of where the company was going and what were to be its goals. We quickly realized that although Tom and Patrick had a successful collaboration at the university, they were really working on two different aspects of biotechnology. The microfluidic cell sorting invented by Soh was potentially useful in diagnostics as well as creating reagents for research, whereas Patrick's invention was directed toward cancer therapeutics. Thus, after a relatively short time, we split the company in two. The therapeutics company carried the CytomX name, while the cell sorting company became Cynvenio Biosystems, Inc. After this was completed, the Board hired Nancy Stagliano as CEO for CytomX and André de Fusco as CEO of Cynvenio.

The CytomX concept is to block the function of an antibody cancer drug by attaching a peptide linker which is in turn attached to a "mask" that would cover the receptor and therefore make the drug inactive. We called that masked antibody drug a "probody." The linker could be cut by a protease (a molecule that cuts proteins). The goal was to utilize a linker that could be cut by a protease that was found specifically in or near the tumor. Because such antibody drugs often have severe side effects, the implied directed delivery would enable the use of higher doses and consequently greater efficacy without the harmful side effects. The initial experiments looked promising, but efforts to obtain venture capital funding were not successful. With diligence, great effort and some reliable scientific progress, Nancy was able to find the necessary funding from Third Rock Ventures under the condition that we move the company to San Francisco, where it would operate under new management. CytomX has demonstrated excellent progress, sufficient to attract interest and co-development with several large drug companies. I anticipate that there will be an IPO or an acquisition of the company when the efficacy of the probody approach is thoroughly tested and proven in the laboratory and in clinical trials. As I write this, CytomX is moving rapidly and successfully toward such a "monetizing" event.

Cynvenio moved to Thousand Oaks, nearer the home of the CEO and in an area where there were many people with biotech experience, principally because of the success of Amgen. The initial microfluidic chips were fabricated in the clean room facility at UCSB with a multi-step process that was both slow and costly. André and the gang at Cynvenio developed a method to literally stamp the microfluidic "chips" from plastic. They refined the instrument design and the chip, and the cell sorting ability improved, resulting in the possibility of isolating and capturing specific cells with high purity. The purity and specificity became sufficiently good that the decision was made to take on the challenge of capturing and analyzing circulating tumor cells (CTCs). The existence of CTCs had been known for more than a century; CTCs are the principal means of metastasis of cancer.

Cancer is a genetic disease; it is caused by a mutation in a specific gene. Different cancers (e.g. breast cancer or colon cancer) have the mutation in a different specific gene. Because the CTCs are sloughed off the tumor into the person's blood, they contain all the DNA mutation information that determines the source of the cancer. The cancer-causing mutation could come from smoking or from other environmental hazards, or a person could be born with that mutation. If one could capture the CTCs and identify the mutation by sequencing the DNA, one would have a "perfect" diagnostic; the ultimate in personalized medicine. Despite efforts by many university groups and many companies, no one had succeeded in capturing CTCs with sufficient purity that the cells could be opened, the DNA extracted, amplified by PCR and sequenced. Cynvenio succeeded in this set of tasks. The polymerase chain reaction (PCR) is a technology in molecular biology used to amplify a single copy or a few copies of a piece of DNA across several orders of magnitude, generating thousands to millions of copies of a particular DNA sequence.

The Cynvenio method utilizes magnetic beads onto which the receptor for a specific protein has been attached. Blood cells which are made in the bone marrow have no surface proteins. Thus any cell in the blood that has such a surface protein is a CTC candidate. Proof that this cell is a cancer cell requires analysis of the DNA and the identification of the mutation.

Cynvenio succeeded in capturing and sequencing CTCs for the first time. In an early trial, 10 tumor cells were spiked into blood from a healthy person. The sample was then put through the Cynvenio instrument and microfluidic chip and nine of the ten cells that had been spiked in were recovered! Because such genetic information had previously been available only through tissue biopsy, the Cynvenio process was called and trademarked as *"Liquid Biopsy."* The patient simply has to give a blood sample as is done in the doctor's office during a routine physical examination. Liquid Biopsy™ is simple, non-invasive, lower cost and does not cause the pain typical of a tissue biopsy.

Cynvenio now has a qualified lab and is carrying out tests for patients. Ruth and I went to the lab for a blood draw. We were informed the following day that we had no CTCs — a simple statement that provided us with that most precious of feelings, peace of mind.

There are more than 2.5 million women in the U.S. alone who have survived breast cancer. But one knows that these women worry every night as they are falling asleep whether it will return in the form of bone cancer (as is sometimes the case). These women no longer have to worry; a simple Liquid Biopsy™ test done periodically will give them peace of mind. Other examples of the utility of Liquid Biopsy include monitoring the efficacy of an expensive drug during therapy, and early detection of cancer prior to the emergence of any symptoms.

The Cynvenio Liquid Biopsy™ process is outlined in the figure below.

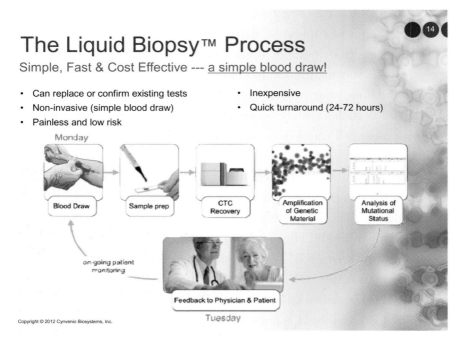

The Liquid Biopsy™ Process
Simple, Fast & Cost Effective --- <u>a simple blood draw!</u>

- Can replace or confirm existing tests
- Non-invasive (simple blood draw)
- Painless and low risk

- Inexpensive
- Quick turnaround (24-72 hours)

Monday

| Blood Draw | Sample prep | CTC Recovery | Amplification of Genetic Material | Analysis of Mutational Status |

on-going patient monitoring

Feedback to Physician & Patient

Tuesday

Copyright © 2012 Cynvenio Biosystems, Inc.

Cynvenio Biosystems is on the pathway toward success. Of all the things that I have done, I am perhaps most proud of Cynvenio. Cynvenio will save the lives of many people and give peace of mind to many others.

20 Chapter

The Delicacy of the Creative Mind

Nearly everyone is familiar with the book by Sylvia Nasar, *A Beautiful Mind*, in which she tells the remarkable story of the schizophrenia and subsequent recovery of John Nash, Jr. Although his journey into mental illness was extreme, the creative mind is a delicate beast. Among creative people, including great scientists and Nobel Laureates, mental illness is not uncommon.

I briefly recounted the story of my close friend, Tony Jensen, in Chapter 10. Tony committed suicide because of the fear that he might lose his creativity as he grew older.

John Robert (Bob) Schrieffer is a great man of science. As a graduate student working with John Bardeen, Schrieffer created the many-body wave function that provided the solution to the remarkable phenomena associated with superconductivity. Superconductivity was discovered by Kamerlingh-Onnes at the University of Leiden (the Netherlands) in 1911 shortly after he succeeded in the liquefaction of helium, thereby making ultra-low temperatures available for scientific studies. He discovered that in some metals and alloys, the resistance to electron flow drops precipitously to zero at a specific transition temperature (different for different metals). For example, for mercury (Hg), the superconducting transition temperature is 4.2K, nearly 269 degrees below the freezing point of water. In the superconducting state, all magnetic fields are expelled from a piece of the superconducting material enabling such a sample to levitate in a magnetic field. Kamerlingh-Onnes received the Nobel Prize in Physics in 1913 for his discoveries.

For nearly 50 years, physicists had been unable to create a theoretical understanding of the origin of these remarkable phenomena. Superconductivity was the BIG problem in solid state physics.

I have emphasized repeatedly that to do great physics, one must have the precious opportunity of working with a great physicist. Schrieffer, of course, understood this and chose to do his thesis research under the direction of John Bardeen. Bardeen, William B. Shockley and Walter H. Brattain were awarded the Nobel Prize in Physics in 1956 for their discovery of the transistor. When Bardeen departed from his home in Champaign-Urbana, Illinois (the University of Illinois), he already knew he would go again in a few years for his contribution to the BCS theory of superconductivity.

Bardeen made a frontal attack on the theory of superconductivity. He attracted Schrieffer (a graduate student) and Leon Cooper (a young post-doctoral fellow) to his group to work on this problem. Cooper made a critically important initial step, but the BCS wave function that provided the deep understanding of the phenomena associated with superconductivity was formulated by Schrieffer. As the story goes, Bob had his critical insight while on the subway in New York during the holiday season of 1956. Thus, Schrieffer became famous at the tender age of 25. This is a heavy load to carry for anyone, especially for someone so young. Bardeen, Cooper and Schrieffer were awarded the Nobel Prize in Physics in 1972 for the BCS theory of superconductivity.

Interest in superconductivity was resurrected in 1986 with the discovery of oxide materials which exhibited the transition to the superconducting state at high temperatures, for example as high as 140K (only 135 degrees Centigrade below the freezing point of water). The discovery of the high temperature superconducting cuprate oxides by Karl Müller and Johannes Bednorz at the IBM Research Laboratory in Zurich in 1986 was a major and wholly unexpected breakthrough. Bednorz and Müller received the Nobel Prize in Physics in 1987. The quest to continue to raise the transition temperature — even perhaps to room temperature — goes on. While it is clear that the basic pairing concept that is fundamental to the Schrieffer (BCS) wave function is valid for these new materials, the mechanism that yields interactions sufficiently strong to enable superconductivity at high temperatures remains an active and controversial area of theoretical research.

Bob had a history of episodes of manic-depression. The success and acclaim that came with the award of the Nobel Prize did not cure him; he had a delicate mind, and was diagnosed as bipolar. I had many experiences with Bob Schrieffer when he was depressed. He looked terrible and was clearly suffering but, remarkably, he still had deep insight into physics. I remember on one such occasion being puzzled by some results that I had obtained. I purposely sought advice from Schrieffer because I thought that thinking about this problem might help him get through this period of depression. I was surprised (amazed!) how quickly he saw through the physics and helped me understand my problem.

He took drugs to help him through these episodes. Although effective at keeping him functional, the side effects caused him to lose his characteristic vigor and enjoyment of life. As I noted earlier, a group of former Penn people who remained close friends after leaving the University of Pennsylvania held a yearly reunion to celebrate New Year's Eve and the coming of the New Year. Typically, at these New Year celebrations, he was quiet and not involved in lively conversations. In December of 2003, I noticed that Bob was unusually outgoing, acting like the "old Bob" again after many years of being in a low and seemingly flat mood because of the anti-depression drugs. I asked his wife, Anna, about this, and she replied with some concern that he had stopped taking his drugs. I was immediately fearful, and as it turned out for good reason. He began writing physics papers with wild claims; he was clearly on a dangerous path.

Early one morning in 2004, he woke his wife, Anna, and told her that he was going on a trip to California. Because he still had strong connections at UC Santa Barbara, she was not concerned. She thought that he was taking a flight, but he drove across the country to Palo Alto with the sole intent of purchasing a new Mercedes convertible. Prior to his departure, he had received a number of speeding tickets in Florida, a number sufficiently large to have his driver license suspended, but he had not yet been formally informed of his loss of driving privileges. As he drove across the country, he accumulated several additional speeding tickets.

He bought the car in Palo Alto and immediately headed south toward Santa Barbara — but he did not complete the journey. He had a high-speed rear-end collision with a van filled with people. He went off the road and broke his knees in the crash. One of the passengers in the van died as a result of the collision. From analysis of the skid marks and the details of the collision, the police report concluded that he had been driving at very high speed, approaching 100 miles per hour. As a result, he was sentenced to one year in state prison. He served his time, but the medical facilities were sorely inadequate. His physical and mental health deteriorated.

Anna took him back to Florida, but was forced to place him in a care facility. He remains there today. This end of life story is particularly sad for such a great man of science. He was my mentor, and my friend. I love him still. But the creative mind is delicate as demonstrated all too clearly by the descent of Bob Schrieffer.

I had only one period in my life that even approached depression. During the stressful period of the TTF-TCNQ controversy, I was disappointed that I had not foreseen the importance of uncorrelated events, but I continued to go forward at full speed to resolve the issues and to eventually get the science right. When the science was understood and the positive aspects of our studies of quasi-one-dimensional systems were clear, I went on with my life. Although disturbed and "down," I did not lose my nerve.

At this point in my life, after 50 years of living a life of science, I've enjoyed success and satisfaction as a scientist and as an entrepreneur. The recognition that came with the award of the Nobel Prize made life even more sweet. Moreover, in the years following the Nobel Prize, I was elected to the National Academy of Sciences and the National Academy of Engineering. I was elected as a foreign member to the Korean National Academy of Sciences, to the Chinese Academy of Sciences, and I was honored with the award of a dozen honorary degrees from universities in the United States, Europe and Asia.

Nevertheless, in September of 2010, with no specific cause, I began to sink into a deep depression. I was hopeless about the future, and almost non-functional as a scientist. I had real thoughts of suicide; I did not want to live on. Ruth saw this happening and, as might be expected, helped me through this terrible period in my life. The first step was to realize that I was not simply "down," but I was clinically depressed. My doctor had me checked into Cottage Hospital for observation. I remained in the "loony bin" for only two days, after which I was released under the care of a psychiatrist, Dr. Robert Nagy. Nagy prescribed anti-depression drugs, several in succession, but with only partial success. These anti-depression drugs serve to make one more or less functional, but they are not a cure, and they do not lead to an end of the depression.

I continued to go to the university, but I was only barely able to continue interacting with my research students and post-doctoral researchers. I did not enjoy the usual pleasures of going to concerts, movies or the theater. To celebrate my 75th birthday, I met my son, Peter, and my grandson, Brett, in Switzerland for a ski week at Verbier. Even that was a bust. I had no energy and did not enjoy the skiing. While they were on the mountain, I spent most of my time sleeping in the chalet.

I felt hopeless and fearful of the future. I could not envision going on in science, but I knew that if I did not do science, I would not survive. What caused this descent into depression? I still do not know. There were issues: my inability to increase the efficiency of the plastic bulk heterojunction solar cells to values above 10%, the impending Konarka bankruptcy, some concerns with a particular set of experiments, but nothing that would justify the depression. Normally, I would have taken on such problems and dealt with them in a straightforward manner.

In September of 2012, David Gross and I performed the staged reading of *Copenhagen* in Brussels at the centennial celebration of the

Solvay Conference of 1912, famous because all the great scientists of the era were there, including Planck, Einstein, Madam Curie and many of their contemporaries. With the help of the drugs from Dr. Nagy and some important encouragement and care from Ruth, I got through it — and actually did well. Until almost the very last moment, however, I was unsure if I could handle the stress of the performance. The day after the performance, Ruth and I flew to New York to celebrate her 75th birthday.

Because she had decided that this was to be her "Diamond Jubilee Birthday," I bought her a very beautiful and very large diamond pendant. We went to our favorite jeweler in Santa Barbara and designed the pendant together in every detail. The center, a large natural clear white diamond, is surrounded by 12 small yellow diamonds. The top of the pendant consists of natural black diamonds through which passed the platinum chain that completed the necklace. It is spectacular, and Ruth is even more beautiful when wearing it.

The celebration was at the restaurant in Manhattan owned by the famous chef Tom Colicchio. Approximately 50 long-time friends and close family members came to celebrate with us. As expected, the food was terrific. We had a group of singers to entertain us, and most important, we were able to share Ruth's celebration with special friends and family that had always been important to us. This celebration was a wonderful affair, worthy of Ruth on this special day.

Somehow, the success of the *Copenhagen* performance in Brussels followed by Ruth's birthday celebration caused the end of my depression. *Copenhagen* and Ruth's 75th birthday party cured me. My vigor and creative spirit returned.

The comparison between my experience and that of my close friend, colleague and mentor Bob Schrieffer is worthy of comment. The creative mind is indeed delicate, but I was able to get through my depression with considerable help from Ruth, my two sons and the expertise of Dr. Nagy. I had lost my nerve despite my preaching many times in these chapters that one must never do so. Bob Schrieffer's story did not end this well, and he will remain in the special care center in Tallahassee for the rest of his life.

I have no insight upon which to give advice on how to avoid the potential problems of the delicate creative mind. I was saved by good luck, hard work, by a loving family and perhaps by my fundamentally optimistic nature. One must continue to be optimistic about the future and take satisfaction from one's accomplishments. One must take each day one at a time and seek to make each day meaningful and satisfying. And clearly — One Must Never Lose One's Nerve!

21 The Joy of Life

I visited the physics department at Princeton a few years ago, and got involved in an unusual lunchtime discussion. The question was quite simple: Would the human race someday become extinct! I honestly had not thought about this subject ever before, and the initial thought in my brain was: "No! This could not happen." Over the course of the next half-hour, however, it became clear that extinction was a definite (and perhaps even likely) possibility.

This lunchtime discussion was abstract and philosophical. It took place long before the current sensitivity to climate change and global warming. When one begins to understand the implications of climate change and global warming, the possibility of human extinction becomes real and a topic to take very seriously.

I have always wondered what it would be like to come back for a visit in — let's say — 300 years. The Standard Model of particle physics will have been unified with the Gravity Theory and will be taught in great universities because this body of knowledge contains deep understanding of all the known forces in Nature. "Dark matter" and "dark energy" will be understood, and their origins will be common knowledge. Because of its inherent beauty and special significance to physics, the BCS theory of superconductivity will still be taught, although probably at the undergraduate level or possibly even in high school. I envision major progress in perhaps the most difficult problem of all: Understanding the workings of the human brain and understanding the subtle problem of the meaning of consciousness. I wonder whether the "plastic electronics revolution" will have become the core of the new modern technology — and certainly hope that it will have positive impact on the well-being of future generations.

I had not conceived of the possibility of human extinction.

What a shame that would be. We have come so far. We have learned so much about the universe that we live in, so much about Nature through our studies of science, so much about biology and the various diseases that limit our lifetimes, so much about the workings of that heavy mass that we carry around inside our head, and the understanding of all these topics will be greatly expanded during the 300 years beyond the time when I have ceased to exist. But, if humans become extinct, all that knowledge would be lost forever. What a shame, what a waste!

Perhaps there is life similar to ours on distant planets and perhaps those intelligent creatures will have developed a parallel understanding of some of these same great questions, although it is unlikely that their biology or their brain function would be similar to ours. Perhaps they already have, but nevertheless the great progress of humankind on Earth would be gone and of no value whatsoever. Each of us thinks about the end of his or her life. But extinction is different — it is truly the end.

I woke up one morning not long ago having had an epiphany. I realized that I had not existed for period of nine billion years since the creation of our universe with the "Big Bang." Beyond a few more years, I will again not exist for billions of years. The meaning of life is clear to me and to all who give it any serious thought. Life gives us the opportunity to continue to understand more deeply the world we live in, to love, to create new knowledge, to create and listen to great music, to make love, to play with our children and grandchildren and watch them grow, to build a model civilization, to love, and to experience the wonder of the sunrise and the sunset — every day.

These thoughts caused me to consider the meaning of death. People often speak of death as if it were a state: "A person is no longer alive but dead." I think that is not the correct way to think about death. Death is the return to non-existence. A religious person would say that when one dies, he returns to God, but I think "non-existence" is correct.

Extinction and the return to non-existence implies the end of all knowledge. Michael Frayn ends his play *Copenhagen* with a short dialogue between Niels Bohr, Werner Heisenberg and Bohr's wife, Margrethe, on this subject:

Bohr:	Before we can lay our hands on anything, our life's over.
Heisenberg:	Before we can glimpse who or what we are, we're gone and laid to dust.
Bohr:	Settled among all the dust we raised.
Margrethe:	And sooner or later there will come a time when all our children are laid to dust, and all our children's children.
Bohr:	When no more decisions, great or small, are ever made again. When there's <u>no more uncertainty</u> because there's <u>no more knowledge</u>.
Margrethe:	And when all our eyes are closed, when even the ghosts have gone, what will be left of our beloved world? Our ruined and dishonored and beloved world?

But Frayn omitted in this short discourse perhaps the most important loss — the loss of Love.

So, we have a responsibility, a great responsibility to make certain that extinction of the human race does not occur, that the wonderful aspects of life that each us enjoys and benefits from continues indefinitely. Why not? We humans are sufficiently intelligent, sufficiently creative and sufficiently innovative that we can solve even the most difficult problems that future generations will face. We can harness the power of the sun for our energy needs. We can learn how to sufficiently coax the weather patterns such that there will be water for agriculture. We can limit population growth — or better — decrease the birthrate and thereby decrease the number of people on Earth to a level consistent with the sustainable resources of the Earth. We are capable of creating a utopian society where people can live their wonderful lives and continue to understand more deeply the world we live in, to love, to create new knowledge, to create and listen to great music, to make love, to play with their children and grandchildren and watch them grow, to build a model civilization, to love, and to experience the wonder of the sunrise and the sunset — every day.

I started my scientific career as a physicist, I was educated as a physicist; people even tell me that I look like a physicist. Certainly, I think like a physicist. Then, on October 13, 2000, I became a chemist. This life-changing event resulted in opportunities for me that most people are never able to enjoy. I have been invited to lecture about my recent research results on every continent (except Antarctica) and I have done so. I have been invited to lecture on my recent research results in every state within the United States (again, I have done so). Often Ruth accompanies me, especially to a destination of particular interest. We are always treated well, in Asia almost as royalty. These invitations continue unabated even now, 14 years after I received the Prize, even though there is a new "class" of Nobel Laureates every year. I have received more than a dozen honorary degrees from universities in the United States, Europe, Asia and Israel. In addition, I was elected as a foreign member of the Chinese Academy of Sciences and a foreign member of the Korean Academy of Sciences. The Heeger Center for Advanced Materials was created at the Gwangju Institute of Science and Technology in Korea in 2005. The Heeger Beijing Research and Development Center was created at Beihang University in 2013; I serve as the Chief Scientist of this Center. The Alan Heeger Technology Center was recently formed by a company in Yan Xi (near Shanghai); the Jiangsu Juhend New Materials Co., Ltd. Thus, the walls of my offices are covered with documents of honor. More important, I have much more science to look forward to in the coming years.

Despite these many honors, there is always some event that serves to bring one back "down to earth." One such event occurred a few years before the award of the Nobel Prize. Ruth and I were returning from a trip to Asia (from Japan if I remember correctly). We arrived at LAX and waited our turn in the immigration line. When at last we reached the front of the line, the officer beckoned us forward. He studied my passport, put it through his computer and then stopped and looked at me. "What do you do?" he asked. I responded that I was a scientist. "Are you a famous scientist?" he asked. After a moment of thought, I said: "Well, maybe, sort of, at least in my own field of research." He said: "Congratulations! Might I have your autograph?" I smiled and said: "Of course." Then he returned my landing card to me and simply stated that I had not signed at the bottom of the form! An immigration officer with a sense of humor — I have been watching for him ever since, hoping that I could have the opportunity to state that: "Yes, I am a famous scientist. I was awarded the Nobel Prize." But alas, I have never again had the good fortune of being processed by such a good-natured immigration officer. That was indeed a unique welcome home to the United States.

I have succeeded as a scientist beyond my wildest dreams, and fortunately, even as I approach the venerable age of 80, I continue to be the leader of the science of plastic electronics with continuing production of important scientific publications by my research group, publications that are highly cited by my scientific colleagues in the field of plastic electronics, implying that my current research has significant impact on the research plans and progress of scientists all over the world.

I have loved, and I have shared it all with Ruth. We have seen our sons grow to maturity and become successful and respected scientists. We have seen our grandchildren emerge as the next generation with the best part of their lives in front of them.

Equally important, I have created new knowledge, and carried that knowledge forward toward practical applications. I have created and co-founded several new companies with new technologies that range from organic light-emitting diodes (UNIAX Corp.) to metal-oxide backplanes for the next generation of OLED displays (CBRITE Inc.), to the creation and participation in the improvement of the efficiency of the conversion of sunlight into electricity and the demonstration of large-scale roll-to-roll manufacturing of "plastic" solar cells (Konarka Technologies) to the development of the ultimate test for cancer by Cynvenio Inc., the Liquid Biopsy™ diagnostic based directly upon personalized medicine; new technologies that I hope will be of significant benefit to mankind.

I have enjoyed music, and I have been deeply involved with theater. When I give a public lecture about my life to a group of university students, the response is typically remarkable and, to me, somewhat surprising. After all, I am just a kid from Nebraska! Nevertheless, somehow the multiple aspects of my life have come together in a manner that inspires young people.

Fortunately, my health remains good and I continue to enjoy my favorite sport, downhill skiing. The photo below shows me on the ski slope at Park City, Utah, at age 79.

What more could one ask out of life? Well, perhaps one more thing — to ski with my sons Peter and David and my grandchildren Brett, Jordan, Julia and Alice on my 85th birthday in the Back Bowls of Vail under bright blue skies on a sunny winter day!

Alan with grandsons Brett (*right*) and Jordan (*left*).

Progress in science will go on long after I am gone. Thus, perhaps I should conclude with some advice for young scientists who aspire to scientific accomplishments that could have sufficient impact to generate Nominations and eventually to result in the award of a future Nobel Prize.

- ***Cherish creativity!***
- ***Be bold, and have the courage and the audacity to seek to discover!***
- ***Never Lose Your Nerve.***
- ***Remember that creativity and discovery necessarily involve Risk. Dealing with that risk is part of the thrill and satisfaction of living a life in science.***

Final Comments: Mentoring Young Scientists

Malcolm Gladwell, in his book *Outliers*, makes a convincing argument that 10,000 hours of dedicated, hard work are required for an individual to acquire the expertise necessary for success in his or her field. In the academic world, this 10,000 hours of dedicated, hard work occurs first during the period of graduate studies and research as the graduate student works toward fulfilling the requirements for the PhD degree. The period of post-doctoral research is similarly demanding.

Mentoring Post-doctoral Researchers and Graduate Students

Post-doctoral researchers come to me with strong recommendations from their thesis advisors, generally scientists whom I know and trust. Moreover, they have significant scientific accomplishments on the record, publications in major journals and the experience of giving presentations at scientific conferences. As experimental scientists, they have mastered difficult and demanding scientific instrumentation. Thus, they are well educated, and have already demonstrated a relatively advanced level of scientific maturity. So — what can I teach them?

My approach to the mentoring of post-doctoral researchers has focused on helping them mature in multiple ways. The first step is to break down the barrier that they perceive separates them from me as a result of my experience and accomplishments: I insist that they call me by my first name, Alan. This is often surprisingly difficult for them, but a very valuable first step forward in our relationship.

During their period as a post-doc, they should be introduced to a new area of research or at least a new experimental technique. In fact,

I typically try to have the individual work on more than one problem during the period that he/she is with me. The post-doctoral period should take the individual well beyond the PhD and put him or her on the path to becoming a mature young scientist. Often, I simply sit for relatively long periods with a one or two post-docs at the table in my office, and I try to convey to them what is the essential missing point, e.g. why is the fill factor in a bulk heterojunction solar cell so low. Typically, I do not know the answer to the question under discussion, but just as often we, together, come up with interesting ideas that must be explored.

The post-doc must learn to communicate his/her scientific results with clarity, in one-on-one discussion, in public presentations, and in discussions with colleagues. I especially enjoy meeting with them one-on-one to discuss their scientific progress. In such discussions, they begin to learn how *I* think about science, what *I* think is a "good problem" to study, and begin to understand why *I* think that is a "good problem." Acquiring "good taste" in the choice of problems is one of the most important aspects of science that I can teach these young scientists.

It is essential that I continually hammer into them the need to communicate clearly in oral presentations, in one-on-one discussions, and of course in the writing of the manuscripts in which they report their work. I am often surprised how poorly they do initially in writing a good manuscript for publication — even if they have spectacular results to present. Thus, I work closely with the individual and put significant effort into working with him/her during the writing stage. In a typical case, there are many, many drafts before the final manuscript is considered ready to be submitted. The process is not unlike the decanting of a great wine. One must remove the "impurities" and the misleading or unclear statements. The impact of a scientific publication is directly related to the clarity of the presentation. A publication should teach the readers something new and, hopefully, something of value. Nevertheless, despite this serious effort, we are often subsequently confronted with comments from reviewers that demonstrate either that we have missed something important or that we have simply not, in fact, been sufficiently clear. Revisions are made and the paper is improved. The system works; the end product is certainly clearer and more likely to be correct.

The oral presentation is another essential form of communication. The young scientist must learn to present his/her work in such a manner that it will have impact on the listener rather than simply presenting the facts. I find that even the very best young post-docs often give unsatisfactory oral presentations. I teach them how to organize their thoughts. I emphasize that I understand that presenting one's work is not simple; it is in fact a very scary thing to do. One is putting himself/herself on the line in public and taking a risk in doing so. This takes courage, and I want them to understand that this courage is essential; one must never lose one's nerve. They must therefore be confident and as certain as possible that they deeply understand what they are saying and that what they are saying is true. They must learn the various tricks and methods involved in giving a good oral presentation (for example a PowerPoint presentation at a conference). Finally, they must learn to directly confront an issue with confidence when they are either challenged or simply asked a question. The latter is very important. Typically, a young student or post-doc will instinctually defer to the more experienced scientist who made the challenge or asked the question. In fact, the young scientist who is presenting the work almost always knows more about his/her results than the one who asks the question. They must learn to be strong and not "cave in" just because some famous scientist challenges their conclusions!

Traditionally, research groups directed by a professor have weekly group meetings. The idea is to help the other members of the research group keep up with the broader activity, and to help the professor monitor the progress of the person making the presentation. In the early years of my career, that was precisely my mode of operation. After moving to Santa Barbara, I gradually moved away from this practice. The post-doc or graduate student received more attention from one-on-one discussion or from small groups sitting together around the table in my office. I realized, however, that the group meeting served another very different purpose. It served to teach the individual to make a strong and effective presentation that would have impact on the listener. Thus, after doing without the weekly group meeting for several years, I recently returned to this more traditional mode. It works well and serves all the purposes enumerated above. Moreover, I find that I am often surprised by some aspect of new data that I had missed in the individual discussion or by the analysis of data presented. So I am once again convinced of the value of the group meeting. We meet once a week for approximately one-to-two hours, typically with presentations from one individual. I often interrupt; sometimes about a point of science but more often about a point of presentation — how to make the presentation more clear.

In my research group, we work in a very interdisciplinary environment — a unique mixture of physics, chemistry and materials science. In such an environment, the post-doc or graduate student can learn a great deal from his colleagues. Thus, I encourage open discussion, and when I observe such a discussion ongoing, I try to stay out of it and give them a chance to understand one another.

Again and again I "preach" to them of the deep satisfaction of having an abstract idea, then, after doing the required experiments that demonstrate that the idea was correct, discovering that Nature actually behaves as conceived in the mind of the scientist. My goal is that they will mature to be as fortunate as I have been during my life as a scientist. Discovery is the heart of science. When one of my graduate students or post-docs makes a discovery (when it all comes together — even something that is not earth shaking) I emphasize that they should remember that day for it is a special and unique kind of experience that only a scientist can truly understand and appreciate.

I am convinced that one can only learn to do great science from a great scientist. I therefore consider it a pleasure and an honor to try to pass along what I have learned from the great scientists with whom I have worked.

Mentoring Graduate Students

Choosing good graduate students requires a subtle talent. One is trying to ascertain whether or not this young, inexperienced person has what it takes to become a successful scientist. Intuition plays a strong role, but sometimes surprising facts come out in the initial discussion between the would-be student and the professor. As an example, in the early 1980s, Duncan McBranch approached me and expressed interest in joining my research group. He had a good undergraduate record, but I wanted some insight into who he really was. I asked him what he had done the previous summer before entering graduate school. He responded that he had spent the summer on his grandfather's farm. I was somewhat disappointed, but pushed the conversation forward. What, I asked, had he done during the summer on the farm? He responded that he found an old car in his grandfather's garage. The car had not been used in years, and it was completely non-functional. Duncan took the machine apart, fixed and cleaned the components, etc., and lo and behold was now driving it around campus! This was clear evidence to me of the kind of talent that could make a great experimental physicist. I offered him a position immediately, and I was never sorry that I did so. He did beautiful work during his thesis research, and he is now a senior staff member at the Los Alamos National Laboratory.

When mentoring graduate students, the goals are the same as those discussed above for post-doctoral researchers. The difference, however, is that these younger people have less experience. They have often not been involved in research or only peripherally so by helping during the limited time that is available to them during their undergraduate studies. In the early days of my life as a professor, I always worked directly with the graduate student in the laboratory. I typically went back to the lab in the evening after dinner to continue this close interaction.

During recent years, I have removed myself from the laboratory. In fact the students and post-docs will not allow me to do experiments either by myself or in collaboration with one of them; I am no longer competent to do the sophisticated, computer-controlled experiments that are the heart of our scientific effort. Thus, I often start a new student by having him or her work closely with and learn from a more advanced post-doc. This is very effective; they begin to do original work in a narrow area remarkably soon after starting their graduate research. I attempt to give them breadth by encouraging collaboration and forcing them to work on several projects as part of their thesis research. The good graduate student quickly becomes part of the group and is respected by his/her colleagues for his contributions. Certainly mistakes are made, but that is how we learn. By the end of his/her thesis research, I often find that the mature graduate student is more creative and more adept in the laboratory than the typical post-doc.

The learning process does not stop with the award of the PhD or after a period of post-doctoral research. When one changes his/her field of research, another 10,000 hours is required. I have changed the focus of my research several times. You might think that this is an unpleasant chore. It is not; the process of expanding one's knowledge to such a high level in a new area is exciting and continues to be exciting as one proceeds through a lifetime of science.

The results obtained from experiments, as described in this book, were obtained through the hard work and many hours, typically more than 10,000, in the laboratory by the graduate students and post-doctoral researchers who really did the experiments. I am, therefore, deeply in debt to the many graduate students and post-docs who have worked with me over the past half-century. I have listed their names, country of origin and the mid-year of their stay with me in the pages that follow.

I have also greatly benefited from the Visiting Scientists who spent time in Santa Barbara collaborating with me and helping my students and post-docs while on sabbatical leave. The names of these colleagues are also listed.

These lists comprise a large number of people, far too many for me to describe their individual accomplishments in any detail. Thus, I will make only a few comments about specific individuals.

Dr. Yong Cao is a great scientist. He has a remarkable ability to obtain critical data to prove a point. He came to my lab in 1987 and was the first employee at UNIAX Inc. After leaving Santa Barbara to return to China in 1999, he created an Institute for the Study of Semiconducting and Metallic Polymers in the South China University of Technology in Guangzhou. He is now a recognized world leader and was elected to the prestigious Chinese Academy of Sciences.

Dr. Daniel Moses moved with me from Penn to Santa Barbara, where he became a Research Professor. He continues to love to be in the laboratory doing experiments. I thank Dan especially for dragging me into the realm of ultrafast pulsed laser measurements.

Shahab Etemad did many of the critical experiments that proved the reality of solitons in polyacetylene as proposed by the theoretical work of Su, Schrieffer and Heeger. He went on to a successful career as a member of the technical staff at Bell Laboratories (subsequently Lucent Technology).

Satish Khanna lived through the TTF-TCNQ era and made important contributions to resolving the controversial issues. He went on to a career at the Jet Propulsion Laboratory, where he rose to the level of Chief Technologist in the senior administration of the JPL.

Paul Chaikin held professorships in physics at UCLA, Penn, Princeton and now at NYU, where he created the Soft Condensed Matter Program. His book on condensed matter physics, co-authored with Tom Lubensky, filled a major hole in the textbooks available to physicists at every level. Paul is brilliant. Every time I have a chance to sit down and talk to him, I leave with a smile on my face because of some new insight from Paul.

Eitan Ehrenfreund became Professor and later Head of the Physics Department at the Technion in Israel. We have continued to collaborate over the many years since he was with me as a post-doctoral researcher.

Kwanghee Lee is Distinguished Professor of Material Science and Engineering at the Gwangju Institute of Technology in Korea, where he founded and serves as Director of the Heeger Center for Advanced Materials. I have enjoyed many years of continuous collaboration with Kwanghee. He received the Korean Science and Technology Medal in 2013.

Arthur Epstein had an outstanding career at the Ohio State University, where he became Distinguished Professor of Physics and Chemistry.

Cuiying Yang did beautiful structural studies on conducting polymers. Her paper on the structure of stretch-oriented polyethylene containing a modest fraction of MEH-PPV is a classic.

Tom Hagler's thesis carefully analyzed the stretch-oriented samples of MEH-PPV in polyethylene. He demonstrated the large anisotropy in the transport, electro-absorption and luminescence. His results enabled the direct measurement of the exciton binding energy as only 0.05 eV.

Gang Yu came to me as a graduate student, supposedly for one year after which he would return to China. He stayed and completed a beautiful thesis. He then worked with me at UNIAX. His contributions at UNIAX created much of the value of the company when acquired by DuPont in the year 2000. Gang, Boo Nilsson and I co-founded CBRITE Inc. in Santa Barbara in 2004. Gang Yu serves as CTO of CBRITE Inc., where he is focused on creating a new technology for backplanes for LCD and OLED displays.

Xiong Gong came to my research group as a post-doc in 2001. He is an outstanding researcher. He has deep insight into what direction to go to solve a problem, but he nevertheless tries all possible alternatives. Xiong wanted to be a professor, but was hindered in finding a faculty position by his difficult accent. Even I, after many years, had difficulty in understanding him. However, he never gave up and finally succeeded in getting appointed as Assistant Professor at the University of Akron. I was pleased but not surprised that he quickly built a strong research group and began to contribute important papers to the literature. I was surprised, however, that he received the highest teaching ranking of anyone in his department. He told his students that he had a difficult speech problem, that he would attempt to write everything clearly and in detail on the blackboard, and that if there were ever any questions, that they should not hesitate to ask. His success as a classroom teacher is a tribute to his strength and motivation as a scientist.

Michael McGehee expressed his desire to be a professor when I first met him. After completing his PhD, he went directly to Stanford University as an Assistant Professor of Materials Science without the typical experience of a period as a post-doc. He is now Full Professor of Materials Science and a major figure in the science of semiconducting polymers.

Jun Gao is Professor of Physics at Queen's University in Canada. His studies of the light-emitting electrochemical cell provided a definitive answer to the mechanism of operation.

Sarah Cowan did beautiful work on bulk heterojunction solar cells. Her discoveries cleared up many of the mysteries and controversies that plagued the field. She is now Staff Scientist at the National Renewable Energy Laboratory (NREL).

Curtis R. Fincher became Senior Scientist at DuPont and had the lead responsibility for their OLED display project.

Aharon Kapitulnik is Professor of Physics at Stanford University. He made many contributions to the understanding of high-temperature superconductivity and has gone to carry out a series of cleverly designed experiments that have established him as one of the world leaders in experimental condensed matter physics.

Serdar Sariciftci is Professor of Physical Chemistry at the Johannes Kepler University in Linz, Austria. He is the only person of Turkish decent that has been appointed to a Full C-4 Professorship in Austria or Germany. He was awarded the prestigious Wittgenstein Prize in 2012.

Anshuman Roy started his company, Rhombus, after leaving UC Santa Barbara. At Rhombus, he developed a remarkably sensitive neutron detector. He is on the pathway to major success having developed relationships with the FBI, the CIA and Homeland Security. The ability to detect neutrons with high sensitivity will protect Americans from the import of fissile material that might have been brought in by terrorists for use in nuclear weapons.

Loren Kaake has accepted a faculty position at Simon Fraser University in Vancouver beginning in the Fall of 2014. Loren is brilliant, and I look forward to watching his career blossom. His insight into the origin of the ultrafast charge transfer in bulk heterojunction solar cells provided the route to a deeper understanding of this remarkable phenomenon.

Among the Visiting Scientists, I note that Richard Friend, now Sir Richard Friend, holds the Cavendish Chair of Physics at Cambridge University. Although Richard and I often disagree on aspects of the physics of semiconducting polymers, the dialogue in the scientific literature has resulted again and again in deeper understanding of the basic physics.

There are many more on the following lists who have had successful careers. Indeed, among the former graduate students and post-docs, there are now 50 professors at universities in the United States, Europe and Asia.

I am grateful to all for their contributions to my scientific life. Without their hard work and deep insight, there would not have been a Nobel Prize for Alan Heeger.

Last, but certainly not least, I express my gratitude to my colleagues, the synthetic chemists who design and synthesize the semiconducting and metallic polymers that we and many others in the field use for their studies.

These include, of course, my two colleagues with whom I shared the Nobel Prize, Professor Alan MacDiarmid (now deceased) and Professor Hideki Shirakawa.

After moving to Santa Barbara, I have collaborated with three great synthetic chemists, Professor Fred Wudl, Professor Gui Bazan and Professor Mario Leclerc (Université Laval, Quebec). Each of these men has remarkable talent for the design of new materials (new molecular structures) with novel properties and skill for completing their synthesis. Without their insight and talent, the field of semiconducting and metallic polymers would have suffered, and I would have been left behind in the world of materials physics.

Graduate students who did research in my group to fulfill PhD requirements (total 75)

Truman G. Blocker III (USA, 1962)

Theodore W. Houston (USA, 1962)

Alan P. Klein (USA, 1963)

Gary Gladstone (USA, 1967)

Arthur J. Epstein (USA, 1971)

Shahab Etemad (Iran, 1971)

Eugene F. Rybaczewski (USA, 1972, Deceased)

Lawrence B. Coleman (USA, 1972)

Satish Khanna (India, 1972)

Marshall J. Cohen (USA, 1972)

T. S. Wei (Taiwan, 1973)

Arthur A. Bright (USA, 1973)

William J. Gunning (USA, 1975)

Curtis R. Fincher, Jr. (USA, 1976)

J. Campbell Scott (Scotland, 1976)

Richard Spal (USA, 1977)

Dale. L. Peebles (USA, 1977)

L. S. Smith (USA, 1977)

James Kaufer (USA, 1978)

Yung Woo Park (Korea, 1978)

Laurie Lauchlan (USA, 1979)

Adam Pron (Poland, 1980)

Avi Feldblom (USA, 1981)

K. B. Lee (Korea, 1981)

C. E. Chen (Taiwan, 1981)

Jane D. Flood (USA, 1982)

J. H. Kaufman (USA, 1982)

Michael Sinclair (USA, 1985)

Nick Colaneri (USA, 1985)

Francisco Moraes (Brazil, 1985)

Howard Shaffer (USA, 1984)

R. Zacher (USA, 1985)

Jun Chen (China, 1985)

Michael J. Nowak (USA, 1986)

Daniel Spiegel (USA, 1986)

Michael Winokur (USA, 1987)

Neal Basescu (USA, 1987)

David Braun (USA, 1987)

Steve D. Phillips (USA, 1988)

C. M. Foster (USA, 1988)

Duncan McBranch (USA, 1988)

K. Pakbaz (Iran, 1990)

Gang Yu (China, 1988)

Karl F. Voss (Germany, 1990)

Tom W. Hagler (USA, 1990)

Laura Smilowitz (USA, 1991)

Andrew Hays (USA, 1991)

D. Chen (Taiwan, 1992)

C. Halvorson (USA, 1992)

Connie L. Gettinger (USA, 1993)

Kwanghee Lee (Korea, 1993)

K. Vakiparta (Finland, 1994)

B. Kraabel (USA, 1994)

John McElvain (USA, 1995)

Dan Vacar (USA, 1996)

F. Hide (Japan, 1996)

Cuiying Yang (1955–1995)

E. K. Miller (USA, 1996)

Michael McGehee (USA, 1998)

R. Gupta (India, 1999)

Cesare Soci (Italy, 2001)

James Swensen (USA, 2003)

Jun Gao (China, 2001)

W. Ma (China, 2004)

Jonathan Yuen (USA, 2006)

Nelson E. Coates (USA, 2007)

J. S. Moon (Korea, 2008)

P. Ledochowitsch (Germany, 2008) — Undergraduate

Sarah Cowan (USA, 2009)

J. H. Lee (Korea, 2010)

J. Granstom (Sweden, 2010)

G. Rowell (USA, 2010) — Undergraduate

C. J. Takacs (USA, 2010)

Bang-Yu Hsu (Taiwan, 2010)

Jason Seifter (USA, 2011)

Distribution:
USA: 48, Korea: 5, Iran: 2, Europe: 7, Japan: 1, China: 2, Taiwan: 4, India: 1, Brazil: 1

Post-doctoral researchers (total 94)

Dr. Sunil Ghosh (India, 1982)

Dr. John M. Bretell (Australia, 1967)

Dr. L. B. Welch (USA, 1968)

Dr. Anthony Garito (USA, 1972)

Dr. Eitan Ehrenfreund (Israel, 1972)

Dr. Paul M. Chaikin (USA, 1972)

Dr. Joel Cohen (USA, 1973)

Dr. Daniel Sandman (USA, 1973)

Dr. Fred Yamagishi (USA, 1973)

Dr. David B. Tanner (USA, 1974)

Dr. C. S. Jacobsen (Denmark, 1974)

Dr. C. K. Chiang (Taiwan, 1976)

Dr. Paul Newman (USA, 1976)

Dr. Arnold Denenstein (USA, 1977, Deceased)

Dr. B. R. Weinberger (USA, 1979)

Dr. D. P. Chakraborty (India, 1978)

Dr. Masaru Ozaki (Japan, 1979)

Dr. Daniel Moses (Israel, 1979)

Dr. S. Ikehata (Japan, 1980)

Dr. T. C. Chung (Taiwan, 1981)

Dr. S. Etemad (Iran, 1981)

Dr. T. Mitani (Japan, 1981)

Dr. K. Kaneto (Japan, 1982)

Dr. Graciella Blanchet (Argentina, 1983)

Dr. K. C. Lim (Taiwan, 1983)

Dr. Curtis R. Fincher (1983)

Dr. S. Casalnuovo (USA, 1984)

Dr. Aharon Kapitulnik (Israel, 1984)

Dr. R. M. Boysel (USA, 1985)

Dr. S. Hotta (Japan, 1987)

Dr. S. D. D. V. Rughooputh (Mauritius, 1987)

Dr. Yong Cao (China, 1987)

Dr. D. Mihailovic (Yugoslavia, 1990)

Dr. S. Tokito (Japan, 1990)

Dr. H. Tomozawa (Japan, 1987)

Dr. G. Gustafsson (Sweden, 1992)

Dr. N. S. Sariciftci (Turkey, 1992)

Dr. Reghu Menon (India, 1992)

Dr. C. Zhang (China, 1993)

Dr. C. O. Yoon (Korea, 1993)

Dr. Yang Yang (Taiwan, 1994)

Dr. Q. Pei (China, 1995)

Dr. Ulli Lemmer (Germany, 1996)

Dr. Benjamin Schwartz (USA, 1996)

Dr. Maria A. Diaz-Garcia (Spain, 1996)

Dr. D. J. Dick (USA, 1996)

Dr. B. Volodin (France, 1997)

Dr. Mats. R. Andersson (Sweden, 1996)

Dr. Yuval Greenwald (Israel, 1997)

Dr. A. Aleshin (Russia, 1997)

Dr. A. Dogariu (USA, 1998)

Dr. J. Pei (China, 1999)

Dr. W. Huang (China, 2000)

Dr. Paulo Miranda (Brazil, 2001)

Dr. C. Fan (China, 2002)

Dr. Xiong Gong (China, 2002)

Dr. L. Edman (Finland, 2003)

Dr. Roland Schmechel (Germany, 2003)

Dr. Q.-H. Xu (China, 2004)

Dr. J. Y. Kim (Korea, 2006)

Dr. Y. Xiao (China, 2005)

Dr. Rebecca Lai (China, 2006)

Dr. A. S. Dhoot (UK, 2006)

Dr. Shinuk Cho (Korea, 2006)

Dr. I. W. Hwang (Korea, 2008)

Dr. S. H. Park (Korea, 2007)

Dr. E. B. Namdas (India, Singapore, 2007)

Dr. J. K. Lee (Korea, 2008)

Dr. J. H. Seo (Korea, 2009)

Dr. Anshuman Roy (India, 2009)

Dr. M. Tong (China, 2010)

Dr. F. Xia (China, 2010)

Dr. N. Banerji (Switzerland, 2011)

Dr. Jang Jo (Korea, 2011)

Dr. W. L. Leong (Singapore, 2011)

Dr. Jacek Jasieniak (Australia, 2011)

Dr. Dong Hwan Wang (Korea, 2012)

Dr. H. R. Tseng (Taiwan, 2012)

Dr. Loren Kaake (USA, 2011)

Dr. A. K. K. Kyaw (Singapore, 2012)

Dr. David Wynands (Germany, 2012)

Dr. Yanming Sun (China, 2010)

Dr. Byoung Hoon Lee (Korea, 2014)

Dr. Hyosing Choi (Korea, 2013)

Dr. Vinay Gupta (India 2011)

Dr. Piiotr Hancyk (2014)

Dr. Changmei Zhong (China, 2015)

Dr. Huiqiong Zhou (China, 2013)

Dr. Chan Luo (China, 2012)

Dr. David Vonlanthen (Switzerland, 2014)

Distribution

USA: 15, Europe: 16, Japan: 7, Taiwan: 3
China: 16, India: 5, Australia: 2, Iran: 1
Israel: 3, Brazil: 1

Index